T0290894

Green Productivity and Cleaner Production

Green Productivity and Cleaner Production

A Guidebook for Sustainability

Guttila Yugantha Jayasinghe
Shehani Sharadha Maheepala
Prabuddhi Chathurika Wijekoon

CRC Press
Taylor & Francis Group
Boca Raton London New York

CRC Press is an imprint of the
Taylor & Francis Group, an **informa** business

First edition published 2021
by CRC Press

6000 Broken Sound Parkway NW, Suite 300, Boca Raton, FL 33487-2742
and by CRC Press
2 Park Square, Milton Park, Abingdon, Oxon, OX14 4RN

© 2021 Taylor & Francis Group, LLC

CRC Press is an imprint of Taylor & Francis Group, LLC

Reasonable efforts have been made to publish reliable data and information, but the author and publisher cannot assume responsibility for the validity of all materials or the consequences of their use. The authors and publishers have attempted to trace the copyright holders of all material reproduced in this publication and apologize to copyright holders if permission to publish in this form has not been obtained. If any copyright material has not been acknowledged please write and let us know so we may rectify in any future reprint.

Except as permitted under U.S. Copyright Law, no part of this book may be reprinted, reproduced, transmitted, or utilized in any form by any electronic, mechanical, or other means, now known or hereafter invented, including photocopying, microfilming, and recording, or in any information storage or retrieval system, without written permission from the publishers.

For permission to photocopy or use material electronically from this work, access www.copyright.com or contact the Copyright Clearance Center, Inc. (CCC), 222 Rosewood Drive, Danvers, MA 01923, 978-750-8400. For works that are not available on CCC please contact mpkbookspermissions@tandf.co.uk

Trademark notice: Product or corporate names may be trademarks or registered trademarks, and are used only for identification and explanation without intent to infringe.

ISBN: 978-0-367-53509-4 (hbk)
ISBN: 978-1-003-08226-2 (ebk)

Typeset in Times
by Deanta Global Publishing Services, Chennai, India

Contents

Foreword

Green productivity (GP) and cleaner production (CP) strategies have been developed to simultaneously enhance both productivity and environmental sustainability. The application of green productivity tools and techniques in the appropriate procedure is important to improve the productivity and environmental performance of an organization and to ultimately achieve overall socio-economic development. GP is applicable not only in the manufacturing sector but also in the agricultural and services sectors. GP plays a vital role in addressing the interactions between economic activities and community development while acting on environmental protection and awareness in the public sector. GP and CP address all elements of a production system, including inputs, processes, outputs, and waste (including environmental pollution). It ensures that products or services meet customers' requirements and that productivity is maintained or improved.

This book will encourage readers to realize the mistakes of conventional productivity, and equip them with an understanding of green productivity as one of the initial steps in commencing sustainability. It demonstrates the GP approach and outlines how to achieve sustainability in a methodical way through green productivity. This book will act not only as a textbook and reference material for academics of the sector but also as a tool for initiating sustainability within businesses and communities. It is an effort to describe the concept of green productivity and cleaner production in a convenient manner and is accessible to every level of reader who is interested in the topic. The green productivity methodology, tools, and techniques and cleaner production strategy described in the book enable reader to determine how they can contribute to the greening process.

Guttila Yugantha Jayasinghe
Department of Agricultural Engineering
University of Ruhuna
Sri Lanka

Authors

Guttila Yugantha Jaysinghe has a BSc in Agriculture and an MSc in Applied Microbiology. In addition, he holds an MSc in Environmental Science and Technology. He obtained his doctoral education from Kagoshima University, Japan, and has authored more than 100 scientific communications. He is currently serving as a visiting professor to Melbourne University, Australia, and the Kagoshima University, Japan.

Shehani Sharadha Maheepala obtained a BSc in Green Technology from the Faculty of Agriculture, University of Ruhuna, Sri Lanka.

Prabuddhi Chathurika Wijekoon obtained a BSc in Green Technology from the Faculty of Agriculture, University of Ruhuna, Sri Lanka.

Authors

[illegible] technology. He obtained his doctoral education from [illegible]. He has authored more than 200 [illegible] University [illegible]

[illegible]

[illegible]

1 Basic Approach to Green Productivity

Guttila Yugantha Jayasinghe,
Shehani Sharadha Maheepala, and
Prabuddhi Chathurika Wijekoon

1.1 A BASIC APPROACH TO GREEN PRODUCTIVITY

The continuous growth of the global population has resulted in enormous challenges for the global environment and for nature (Foley *et al.*, 2011). The higher living standards of an escalating population that is equipped with increasing wealth and higher purchasing power has led to increased needs and consumption patterns (Godfray *et al.*, 2010). In order to fulfill these rising needs, production patterns have been developed that focus on higher productivities, as shown in Figure 1.1.

1.2 PRODUCTIVITY

Productivity is conventionally defined as "the ratio between inputs and outputs", Equation (1.1), where human, physical, and financial resources and information are considered inputs for a particular production process, and products, services, and wastes are considered outputs (OECD, 2001; Tangen and Stefan, 2005).

$$\text{Productivity} = \frac{\text{Output}}{\text{Input}} \tag{1.1}$$

Productivity is considered a measure of the ability of a particular organization or person to efficiently utilize available resources to produce the desired output. This is the technical concept of productivity (Chakravorty, 2013).

In terms of economy, productivity can be defined as one's ability to create more value for customers; from the perspective of management, it is equated with efficiency and effectiveness (Chakravorty, 2013).

As a technical concept productivity is viewed in quantitative rather than qualitative terms. It may ultimately result in a productivity improvement with increased outputs of reduced quality. It may even result in labor–management conflicts due to the neglecting of social aspects while increasing labor productivity. With the aim of overcoming these problems, productivity has been defined using an integrated approach.

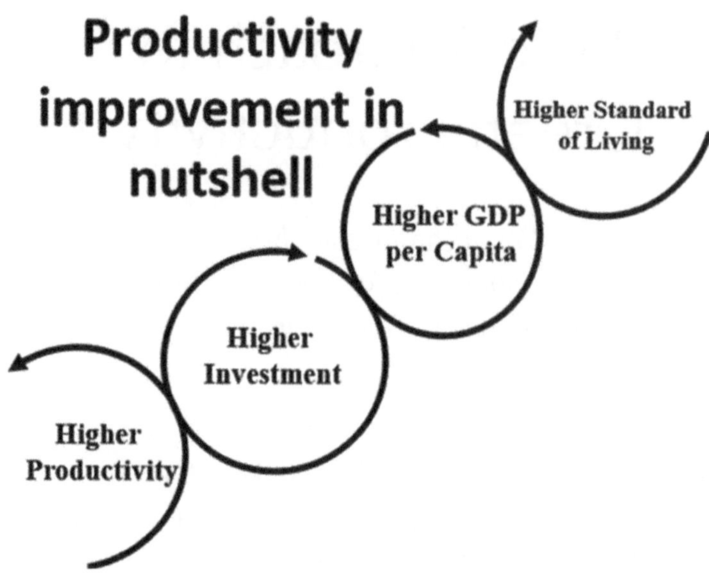

FIGURE 1.1 Productivity improvement in a nutshell.

The integrated approach to productivity is defined by the following equation:

$$\text{Productivity} = \frac{\text{Output} \times \text{Satisfaction}}{\text{Input} \times \text{Sacrifice}} \tag{1.2}$$

Under the integrated approach, productivity can be described as an objective and as a means. As an objective, it is explained through the social concept of productivity, and is considered a mental attitude as well as the above. "It seeks to continually improve what already exists. It is based on the belief that one can do things better today than yesterday and better tomorrow than today." As a means, it is explained through the technical, management, and economic concepts of productivity (APO, 2004).

Productivity is further defined as:

- Doing the right thing in the right way
- Working smarter not working harder
- Increasing efficiency, economy, and effectiveness
- Getting the best out of what is available

The concept of productivity is applicable to any country and to any organization, be it public or private, profit or nonprofit, large or small, manufacturing or service; it is even applicable to the home and to individuals. To derive improved productivities, production processes aim to increase outputs while relatively reducing input consumption. This productivity improvement process should mainly be guided by three principles: to increase employment, maintain labor–management cooperation, and allow fair distribution of the results.

Improved productivity is beneficial for employees in different ways (Harte *et al.*, 2011), such as:

- Increased salaries and other benefits
- Better working conditions (better technology, systems, occupational safety, and health)
- Improved competency and career development
- Pleasant working environment and greater job satisfaction
- Better recognition in society
- Higher living standards

Industrial, agricultural, and service sectors are important in production activities. The conventional model of productivity is based on economy. Production and consumption patterns are driven by the primary aim of achieving economic development. Economic development activities ultimately result in a higher financial productivity that causes fatal issues for the environment. The inputs required for production are obtained from the environment, and the wastes generated in the processes are disposed in the environment also. In this way, the production process has impacts on the natural resource base throughout its life cycle. Productivity improvement increases the rates of resource extraction and waste accumulation on Earth, which subsequently creates a huge stress on the environment (OECD, 2013).

Traditionally, productivity improvement focused predominantly on cost effectiveness through reduced costs. Therefore, in order to improve profitability or organizational effectiveness, the cost reduction approach was used.

With the arrival of quality as a concept and a drive, productivity was measured by comparing the benefits accrued from a quality program (output) with the resources utilized within a particular production system (inputs).

Several programs emerged with the aim of improving productivity by driving the internal organization of an industry. Among these, the Total Quality Management (TQM) approach, together with the Total Productive Maintenance (TPM) system, have become the most popular. TPM mainly addresses equipment maintenance. There is also 5S, the workplace organization method that ensures structured and efficient housekeeping in an organization, and Kaizen, which focuses on continual improvement.

Some productivity improvement practices, including preventive maintenance and good housekeeping, are aimed at reducing the environmental burden a considerable extent. Nevertheless, for total environment management, it is vital to incorporate these productivity improvement programs together. In the Asian region, the need for a viable strategy that integrates the environment into the productivity improvement of the industrial, agricultural, and services sectors has rarely been considered. Meanwhile, most countries are in a rush, pressurized by the following imperatives:

1. To achieve rapid industrial growth with a limited resource base, while ensuring that there is no further deterioration of natural resources
2. To include the environment as a strategic business factor in the international marketplace

3. To increase public awareness and concern for the environment
4. To improve environmental regulation and enforcement

Considering the above aspects, administrations together with regulatory agencies should lead the way forward towards environmental protection, together with the commitment and contribution of the public. Moreover, most enterprises have limited financial resources and, in order to compete in the international market, have to turn around their business strategies as they are complying with international aspects. This process should be used as an opportunity by industrial, agricultural, and services sectors, as well as by policy makers, to steer the regional economy towards sustainable development.

Therefore, according to the integrated approach to productivity, the Earth's environment makes sacrifices in fulfilling the insatiable and rising demands of humanity. Although humans receive a lot of positive resources from nature, they always return the negative to nature.

The result of conventional productivity improvement is the depletion of basic living needs and the disruption of all life on Earth (Jorgenson *et al.*, 2012). The threatening environmental issues are:

- Pollution (air, water, and soil)
- Ozone layer depletion
- Acid rain
- Biodiversity losses
- Desertification
- Environmental health issues
- Climate change
- Natural disasters
- Food security

Humanity has volunteered to suicide itself, threatening not only the current population but also future generations, as the effects of all environmental issues will impact them also. When considering the period of the late 1960s and early 1970s, the communities of western countries had witnessed and been threatened by detrimental environmental issues. However, some environmental problems, including toxic waste contamination, air pollution, and water pollution, were taking place in the Asian region of the world.

With the industrial revolution came the fallout of industrial growth in the 1980s. It had instigated many environmental issues, resulting in huge impacts on the human population. For example, the industrial emissions in the United States had created acid rain in Canada, while forest destruction in Germany was claimed to have been caused by the acidification of water bodies in Scandinavia.

Moreover, most of the developed countries of the world had begun to depend on developing countries for renewable resources, especially fuel and minerals. This built up a trade for these resources, boosting the economies of developing countries. Nevertheless, it originated significant environmental degradation in these countries. Due to the degradation and depletion of natural resources, environmental issues also diffused into those developing countries alongside the economic development. Most Asian countries had been affected in this context. Specifically, industrialization and enhanced energy consumption led to regional problems in Asia, concealed by developing economies. International concern was focused on global warming and ozone layer depletion, ranking them as the detrimental issues of the era.

Global trade diffused further regardless of regional and continental barriers, and there was a greater recognition that environmental degradation does not respect boundaries either. The interaction between nature and humanity had become fatal. Devastating historical incidents provide the best examples for the severity of the issue (Blacksmith, 2007; Bogard, 1987; Newman, 2016):

- Love Canal disaster, New York City (United States) – Toxic waste poisoning
- Minamata disease, Japan – Mercury poisoning
- Bhopal gas tragedy, India – Chemical gas exposure
- Bangladesh wells – Arsenic poisoning of the ground water

In the 1990s, international recognition of these environmental issues increased, leading to the emergence and consideration of those issues at international gatherings and conferences. With this international recognition, sustainable development has necessarily become the need of the hour. All environmental issues and aspects that were previously being discussed and addressed at regional, technical seminars, and debates have increasingly become diplomatic issues for the global community, and have become a vital part of international trade.

It is estimated that the rapidly growing world population will reach 9.8 billion in 2050, 31% higher in comparison to the present figure, and that consumption will increase five-fold. To meet the resource consumption of the population in 2050, resource consumption must be reduced by a factor of between 4 and 10 (UN, 2017).

The need for a sustainable mechanism is further emphasized in the context of:

- Limited resources in the world
- Increasing population
- Changing lifestyle
- Increasing consumption patterns
- Waste culture versus optimum consumption levels

Therefore, achieving sustainability in the patterns of production and consumption has become of greater concern, positioning it as a political and diplomatic issue for the modern world that can influence both business and environment.

1.3 SUSTAINABILITY

With agriculture and the utilization of fossil fuels being, respectively, the first and second transitions to have completely changed the way global society operates, sustainability is considered to be the third transition that humans must address. "Sustainability" simply means the ability to sustain present life, future life, life on earth, and the future of earth.

According to the publication of *Our Common Future* by the Brundtland Commission in 1987, sustainable development is defined as “the development which meets the needs of the present without compromising the ability of future generations to meet their own needs”. Agenda 21, a product of the Rio Declaration on Environment and Development, was adopted by more than 178 governments at the first Rio de Janeiro Earth Summit of 1992 (UN, 1987).

The triple bottom line of sustainability suggests that all three dimensions, including environmental integrity, and social and economic equity, must become the focus of sustainable development. Figure 1.2 shows the triple bottom line of sustainability. The principles of sustainable development should be engaged at all steps of the planning process, as a life cycle approach (Kuhlman, 2010).

Environmental sustainability can be defined as the process of making sure that current processes of interaction with the environment are pursued with the idea of keeping the environment as pristine as naturally possible based on “ideal seeking behavior”. By greening their processes, humanity's systems can be adapted to the sustainability approach as a drive for addressing most global environmental issues. Thus, the concept of greening the productivity process has evolved with the aim of enhancing socio-economic development, which leads to sustained improvement in the quality of human life.

Greening the productivity process focuses on three ecological principles to ensure sustainability in development activities (Lin Sheng *et al.*, 2005). The three principles are:

1. Protection of ecological balance

Pollution and the overexploitation of resources deplete the balance of ecological systems. The output of these activities will exceed the environment's maximum capacity to tolerate waste, and ecological processes will be disrupted. Greening productivity

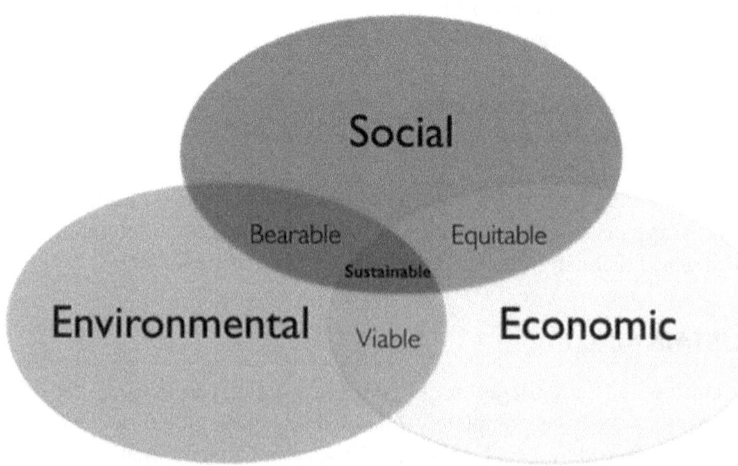

FIGURE 1.2 Triple bottom line of sustainability.

aims to maintain those processes through efficient use of resources and pollution prevention.

2. Sustainable use of natural resources

Natural resources are used extensively for production processes as well as input material. Efficient use of natural resources is emphasized in green productivity as it allows for the regeneration of natural resources.

3. Protect plant and animal species

Plants and animals are essential for the survival of ecological systems. Pollution and overexploitation threaten them due to the habitat degradation that they cause. Green productivity can lead to the survival of those species through efficient resource utilization and pollution reduction.

1.4 SUSTAINABLE DEVELOPMENT GOALS AND GREEN PRODUCTIVITY

The Sustainable Development Goals (SDGs) are a universal set of goals formed in 2015 as agreed by the leaders of 193 countries. The 17 goals were formulated by the United Nations General Assembly in order to frame their agendas and policies over next 15 years (UNDP, 2015). Table 1.1 below shows the SDGs.

The SDGs outline the way forward for development processes that are socially, environmentally, and economically sound. Productivity improvement also can be directed towards a sustainable destination through the integration of the SDGs in production processes. Among the 17 goals, some can be directly incorporated into production processes in order to green the production.

The seventh goal is "affordable and clean energy", which implies enhancing the accessibility of electricity for people worldwide and using renewable and sustainable energy sources (Matte *et al.*, 2015). By achieving this goal in the productivity improvement process, organizations can achieve green productivity. Energy is intensive, but it is a basic requirement for any kind of production processs. In green productivity, the utilization of renewable energy sources in the prodution process is always encouraged. Efficient resource use is a basic concept of green productivity (GP) that further emphasizes the importance of avoiding the use of those resources at a rate higher than it can regenerate.

"Decent work and economic growth" is the eighth SDG and encompasses:

- Inclusive and sustainable economic growth
- Full and reproductive employment
- Decent work for all (UNDP, 2015)

All of these concepts can be met with the green productivity approach. GP focuses on enhancing economic profitability while remaining concerned with the environmental

TABLE 1.1
Sustainable Development Goals

	Sustainable Development Goal	Targets
1	No poverty	End extreme poverty in all forms by 2030.
2	Zero hunger	End hunger, achieve food security and improved nutrition, and promote sustainable agriculture
3	Good health and well-being	Ensure healthy lives and promote well-being for all at all ages
4	Quality education	Ensure inclusive and equitable quality education, and promote lifelong learning opportunities for all
5	Gender equality	Achieve gender equality and empower all women and girls
6	Clean water and sanitation	Ensure availability and sustainable management of water and sanitation for all
7	Affordable and clean energy	Ensure access to affordable, reliable, sustainable, and modern energy for all
8	Decent work and economic growth	Promote sustained, inclusive, and sustainable economic growth, full and productive employment, and decent work for all
9	Industry, innovation, and infrastructure	Build resilient infrastructure, promote inclusive and sustainable industrialization, and foster innovation
10	Reduced inequalities	Reduce inequality within and among countries
11	Sustainable cities and communities	Make cities and human settlements inclusive, safe, resilient, and sustainable
12	Responsible consumption and production	Ensure sustainable consumption and production patterns
13	Climate action	Take urgent action to combat climate change and its impacts
14	Life below water	Conserve and sustainably use the oceans, seas, and marine resources for sustainable development
15	Life on land	Protect, restore, and promote sustainable use of terrestrial ecosystems, sustainably manage forests, combat desertification, halt and reverse land degradation, and halt biodiversity loss
16	Peace, justice, and strong institutions	Promote peaceful and inclusive societies for sustainable development, provide access to justice for all, and build effective, accountable, and inclusive institutions at all levels
17	Partnerships for the goals	Strengthen the means of implementation and revitalize the global partnership for sustainable development

aspects of a production process. The base of GP is built on this concept of sustainable economic growth. Labor productivity is another aspect addressed by GP that also aligns with increasing social sustainability in productivity improvement.

A further SDG that can be fulfilled by GP is the ninth goal, "industry, innovation and infrastructure" (UNDP, 2015). This goal centers on:

- Building resilient infrastructure
- Promoting inclusive and sustainable industrialization
- Fostering innovation

Industrial and other production sectors should be fostered by sustainable technologies and innovations in infrastructure, and by other facilities' development. This will ensure the upholding of the sustainable development concept, together with GP, throughout any production process. The more an organization can invest in innovation and infrastructure, the more it can gain. Thus, productivity can be improved in an environmentally robust manner if innovations and infrastructure are mainly focused on environmental sustainability. GP always encourages innovations in its methodology, through tools and techniques such as brainstorming. Reducing energy and material consumption through technological innovation and design changes will improve the quality of the products, and reduce pollution and costs as well.

"Responsible consumption and production" is the twelfth SDG, and can also be greatly interrelated with GP. The goal emphasizes ensuring sustainable consumption and production patterns (Matte *et al.,* 2015). This is the prime motive of the GP approach, which proves that sustainability is the driving force for GP. Production processes should consume natural resources in a way that preserves them and allows future generations to consume the rest. This efficient use of natural resources is addressed by the principles of both the SDGs and the GP. Green consumerism is thus forcing producers to recognize the environment as an essential component in the production process. Improving the disposal of waste, and reducing and recycling it, also features in both sets of principles. It is therefore considered a main step in conducting production in a more responsible manner.

REFERENCES

APO. "Designing green productivity." *Tech Monitor* July–Aug, 2004: 69–71.

Blacksmith institute. *World's Worst Polluted Places-the Top Ten.* New York: Blacksmith institute, 2007.Viewed 25/05/2018, <http://www.Blacksmithinstitute.org>.

Bogard, W. "Evaluating chemical hazards in the aftermath of the Bhopal tragedy." *International Journal of Mass Emergencies and Disasters* 5(3), 1987: 223–241.

Chakravorty, S.K. "Productivity: A continuously evolving concept." *p-Watch, A Macro View of Productivity Trends,* April–May, 2013: 2–3.

Dunlap, R.E, and Jorgenson, A.K. "Environmental problems." In *Encyclopedia of Globalization*, by George Ritzer, 1–8. Wiley Blackwell, 2012.

Foley, J.A., Ramankutty, N., Brauman, K.A., Cassidy, E.S., Gerber, J.S., Johnston ,M., Mueller, N.D., Connell, C.O., Ray, D.K., West, P.C., Balzer, C., Bennett, E.M., Carpent er,S.R.,Hill, J., Monfreda, C., Polasky, S., Rockstro, J., Sheehan, J., Siebert, S., Tilman, D., and Zaks, D.P.M. "Solutions for a cultivated planet." *Nature* 478(October), 2011: 337–342.

Godfray, H.C.J., Beddington, J.R., Crute, I.R., Haddad, L., Lawrence, D., Muir, J.F., Pretty ,J., Robinson, S., Thomas, S.M., and Toulmin, C. "Food security : The challenge of feeding 9 billion people." *Science* 327(February), 2010: 812–818.

Harte, K., Mahieu, K., Mallett, D., Norvlle, J., and VanderWerf, S. "Improving workplace productivity." *Benefits Quarterly* 3, 2011: 12–26.

Kuhlman, T., Farrington, J. "What is sustainability." *Sustainability* 2(11), 2010: 3436–3448.

Lin Sheng, T., Shamsudin , M.Z., and Ling, L.C. "Sustainable development with green productivity in manufacturing." *IEEE International Symposium*2005: 267–270.

Matte, S., Moyer, L., Kanuri, C., Petretta, D., and Bulger, C. *Getting Started with the Sustainable Development Goals.* Guide, Sustainable Development Solutions Network, 2015.

Newman, R.S., and *Love Canel: A Toxic History from Colonical Times to the Present*, 245–299. Health Study, U.S. Department of Health and Human Services Agency for Toxic Substances and Disease Registry, 2016.
OECD "Measuring Productivity." *Manual*, 2001.
OECD. "Material, resources, productivity and the environment." *Key Findings*, 2013.
Tangen, and Stefan. "Understanding the concept of productivity." *Proceedings of the 7th Asia Pacific Industrial Engineering and Management Systems Conference*, 4–8. Taipei, 2002.
UN. "Report of the World Commission on Environment and Sustainability." 1987.
UN. World population prospects: The 2017 revision, key findings and advanced tables, *Department of Economic and Social Affairs – Population Division*, 2017: 5–10.
UNDP. *Sutainable Development Goals Booklet*, 2015.

2 Green Productivity

Guttila Yugantha Jayasinghe, Shehani Sharadha Maheepala, and Prabuddhi Chathurika Wijekoon

2.1 INTRODUCTION

The Asian Productivity Organization (APO) introduced the concept of Green Productivity (GP) in 1994. The APO defines green productivity as a "strategy for enhancing productivity and environmental performance for overall socio-economic development. It is the application of appropriate productivity and environmental management tools, techniques, and technologies to reduce the environmental impact of an organization's activities, goods, and services". The Manila Declaration on Green Industry in Asia, announced at the APO World Conference on Green Productivity in 1996, declares that environmental protection should be promoted without sacrificing productivity (Johannson, 2005).

GP directs organizations towards the concept of "doing better with less" by using material, resources, and energy more efficiently and sustainably. It helps to reduce the cost of operations through better resource utilization, reduced long-term liabilities, compliance with government regulations, and improved corporate image. Therefore, it has been proven as a practical approach that enables enterprises and communities to enhance their profits and productivity while improving environmental performance (Gandhi, 2013). Figure 2.1 shows how productivity improvement can correlate with sustainable development.

In its formal definition, GP is stated using three key terms or phrases:

- Strategy
- Productivity and environmental performance
- Socio-economic development

GP offers step-by-step guidance with appropriate tools, techniques, and technologies to achieve sustainability in the means of productivity within an organization. GP is based on two key components, including a set of tools used to rationalize the input–process–output model, and a set of defined sustainable practices that will guide the practitioner to achieve the objective of GP (Parasnis, 2003).

GP is applicable not only in the manufacturing sector but also in the agricultural and services sectors. GP plays a vital role in addressing the interactions between economic activities and community development, while acting on environmental protection and the awareness of the public sector (Johannson, 2002). GP addresses all elements of a production system, including inputs, processes, outputs, and waste (including environmental pollution). It ensures that products or services meet

FIGURE 2.1 Green productivity improvement.

customers' requirements and that productivity is maintained or improved. Figure 2.2 illustrates the GP framework.

Thus, it can be further stated that GP is specifically characterized by four distinguishing characteristics as discussed below:

1. Environmental Compliance

   The main focus of GP is environmental protection, the first step of which is compliance. Compliance has become one of the most challenging issues faced by the current industry sector, although it can be achieved through GP practices, as these mainly address pollution prevention and source reduction. Residues will have to be treated and managed with the appropriate use of end-of-pipe treatment techniques. The unique characteristic of GP is that productivity will be improved together with environmental compliance. These practices can subsequently lead to a situation beyond environmental compliance, with the ultimate aim of ensuring quality of life.
2. Productivity Improvement

   Another important aspect of GP, which can be considered as being the other side of the "GP coin", is productivity improvement. The Kaizen approach, which drives towards continuous improvement, can be considered basis of productivity improvement. According to GP strategies, this approach has to be accompanied by environmental protection. The concept of continuous improvement as addressed by Kaizen is achieved by adopting the basic terms of the Plan–Do–Check–Act (PDCA) cycle. This cycle principally aims to ensure not only a productivity improvement as addressed by classical productivity programs, but also an environmental improvement. It is considered a dynamic and iterative process.

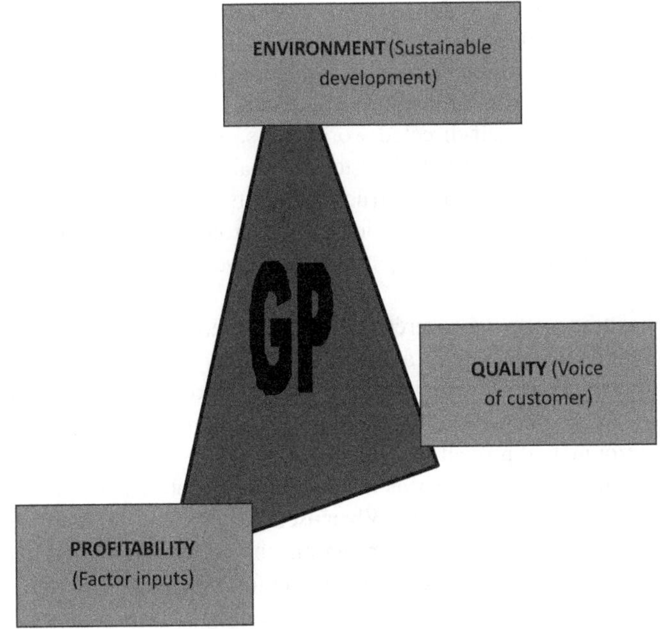

FIGURE 2.2 Green productivity framework.

3. Integrated, People-Based Approach

   One of the strengths of GP is the involvement of workers and employees in its team-based approach. This strength is further enhanced by an improved working environment that ensures worker health and safety, non-discrimination, and minimal social welfare issues in a particular organization. The methodology of GP promotes and involves multi-stakeholder participation throughout its process. This enables a step-by-step sequential approach, a systematic generation of options and solutions, and the contribution of all members in the organization to the GP process. The involvement of all members also ensures transparency and accountability, eliminating conflicts within the organization.

4. Information-Driven Improvement

   Documentation and reporting are the other main strengths of GP, and these are mainly driven by systems such as QMS and EMS. The conveyance of information on what is measured and completed during the process is one of the driving forces of GP. A set of defined GP performance indicators will allow for the continuous measuring and evaluation of the performance of an organization that has established a GP program.

The connection between the conventional productivity improvement model and the GP-driven, environmental protection model is thereby evident. With the aim of achieving their integration, the adoption of novel, innovative, and advanced organizational systems is required. This may allow for the incorporation of green designs, production methodologies, and practices as strategies into agricultural and industrial processes (Chan, 2009).

Advanced systems can be defined as a blend of the technological and organizational alterations of the capital-intensive equipment within a particular organization. These typically include self-directed work teams, worker rotation, and continuous process improvement. This kind of system may also promote interdependent relationships across the organizational structure and its production chain. The relationship can extend beyond the organizational level, from producers to consumers. GP suggests that such an approach helps speed up development towards a sustainable direction.

Traditional trade-offs are overcome by organizations who have responded to increasing competitiveness with adaptations for more effective strategies and by transforming to match the varied nature of competition. GP is accepted as one of the strategies that businesses are looking for to maintain their competitiveness while ensuring environmental protection.

GP can be implemented either as a driver or as a tool that integrates with other management or productivity efforts. When used as a driver, a company-wide administration is required to manage the program and to establish objectives to drive through. When used as a tool, organizations can form GP teams to work on assigned processes (Chan, 2009). GP activities require strong commitment and leadership from top management, as well as strong implementation infrastructure, in order to ensure their effectiveness (Suder, 2006).

An attractive feature of GP is that it can be used as a robust strategy that leads to profitability gains through productivity improvements and enhanced environmental performance. Over-exploitation of resources and the generation of pollution are considered indicative measures of low productivity and poor environmental performance. In many ways, these can be referred to as manufacturing defects that have to be consistently corrected in order to achieve the targets of a particular organization. In order to comply with targets, GP suggests a process of continuous improvement, a strategy that is built upon technical and managerial interventions. Through enhanced resource-use efficiency, GP aims to preserve natural resource "capital", ensuring the conservation of a safe environment for future generations.

With this aim in mind, the first step of the ensuing process is pollution-source reduction, which means identifying ways to prevent pollution, or waste, at its source. At the same time this process seeks to use rationalization and optimization to reduce the level of resource inputs required by the production processes. Reuse, recovery, and recycling possibilities are also investigated in order to minimize the wastes generated.

The next step is to eliminate toxic or hazardous substances by reducing the potential for related pollutions, with the ultimate aim of reducing the life-cycle impact of a product. At this stage, the product life cycle is taken into account, with particular regards to the impact of packaging, or the product's framework of design, for the environment.

As the final step, the generated wastes are treated in their residual forms using adequate end-of-pipe treatment technologies. Suitable solutions have to be found to ensure the organization is complying with regulatory requirements from the perspectives of both the workspace and the receiving environment. Adequate technologies and methods are discussed in Chapter 4, "End-of-Pipe Treatment Techniques". Continuous productivity improvement is ensured through a systematic approach and a sequential methodology built around the GP concept, which will be further discussed in this chapter. Moreover, this approach will enhance the level of environmental protection.

GP is driven by forces which are either external or internal to the organization. Typical external forces include:

- Pressure from regulations, both national and international
- Demands from various stakeholders, such as consumers and suppliers

Regulations may be a strong force for GP. For examples of this, consider the role of complex national regulations and standards, fiscal instruments such as taxes and penalties, and judicial directives. Most national regulations have been formed to reflect international regulatory developments in environmental, natural resource, and ecological protection.

Developing global and industry standards are serving as driving forces for the movement towards more sustainable production processes as facilitated by GP. International conventions such as the Montreal Protocol and the United Nations Framework Convention on Climate Change can be taken into account; moreover, the Responsible Care initiative by the chemical industry, the Marine Stewardship Council established for the food processing sector, the Forest Stewardship Council established for the pulp and paper sector, and more general codes of conduct for environmental and social responsibility can be considered as adequate instances also.

Sometimes, in order to develop economies, these trends may have greater implications for businesses as they can be constrained by some technological and resource factors. The commencement of world markets and the gradually enhancement of globalization has further intensified the pressures on these businesses, which are always in a rush to meet international expectations.

A business must also consider the context of the customer, who seeks quality, cost, reliability and, recently and most importantly, promptness of delivery. However, with improved recognition of the need to integrate environmental requirements into business strategies, together with the pressure from customers' high expectations, suppliers have tended to provide environmentally sound goods and services. Suppliers are expected to obtain standards such as the ISO 14000 and SA 8000 certifications in order to prove the environmental and social performance of their goods and services as required by customers.

GP is an integrated approach that joins several concepts such as cleaner production, energy management, environmental management, and waste management (end-of-pipe treatment). Through the integration of these proven approaches, GP provides a sustainable framework within which to address all the stages of a production process.

2.2 GP METHODOLOGY

The GP methodology is comprised of six steps as depicted in Figure 2.3. These steps can be broken down into 13 tasks, as shown in Table 2.1. This procedure allows enterprises to examine and evaluate their production processes to reduce their environmental impacts and improve product quality (Suder, 2006).

2.2.1 Step 1: Getting Started

2.2.1.1 Task 1: Team Formation

To mark the beginning of the GP methodology, a GP team is formed by guiding members as per conducted analyses. The team can be composed of a core team and multiple sub-teams in large organizations, with the core team supervising the overall GP program while sub-teams assist the core team in specific tasks. Teams should be dynamic and evolving by nature, and led by an open-minded and versatile person (Johannson, 2005).

The team leader should have following qualities:

- An in-depth knowledge of processes and operations in the industry
- Access to all departments
- A healthy relationships with staff
- Managerial qualities

Important qualities to be adopted while working in a group include, cooperation, honesty, active listening, patience, and an openness to feedback.

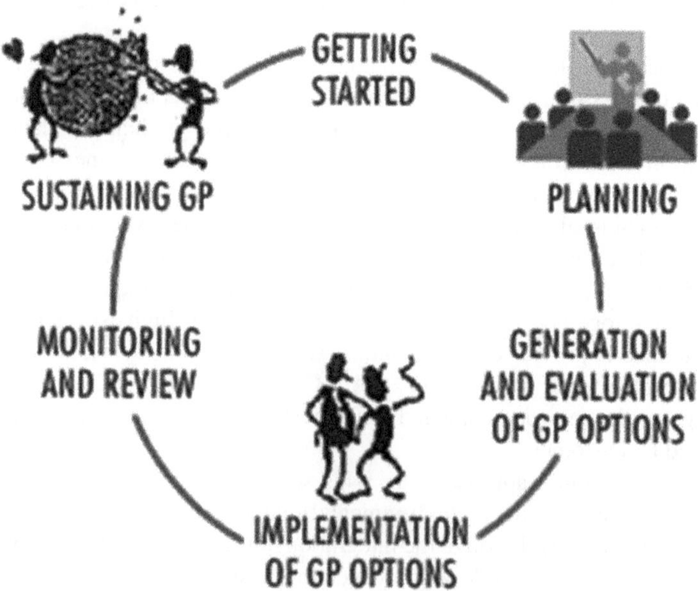

FIGURE 2.3 GP methodology.

TABLE 2.1
The Steps and Tasks of Green Productivity

Step	Tasks
Step 1: Getting started	Task 1: Team formation
	Task 2: Walk-through survey and information collection
Step 2: Planning	Task 3: Identification of problems and causes
	Task 4: Setting objectives and targets
Step 3: Generation, evaluation, and prioritization of GP options	Task 5: Generation of GP options
	Task 6: Screening, evaluation, and prioritization of GP options
Step 4: Implementation of GP options	Task 7: Formulation of GP implementation plan
	Task 8: Implementation of selected option
	Task 9: Training, awareness building, and developing competence
Step 5: Monitoring and review	Task 10: Monitoring and evaluation of results
	Task 11: Management review
Step 6: Sustaining GP	Task 12: Incorporation of changes into organizational system of management
	Task 13: Identification of new/additional problem areas for continuous improvement

2.2.1.2 Task 2: Walk-Through Survey and Information Collection

In this task, the team walks through the site, conducts a survey on the facility, and records their first impressions. This allows the team to gather baseline data and to identify areas where there may be a problem (Johannson, 2002).

Before conducting a walk-through survey, the following information should be collected for the benefit of the GP team members:

- A layout showing process equipment, utilities, storage areas, and offices
- A process flow diagram (at least on the block level) and the date of last update of the diagram
- A layout of water supply lines, drainage channels, steam lines, etc., as applicable
- Production-related information in terms of scheduling

During the walk-through survey the GP team is able to validate and update existing information such as process flow diagrams, layouts, etc. (APO, 2004).

Data collection can be carried out by reviewing existing documents regarding organizational, financial, environmental, service, and production activities:

(a) Organizational information
  - Organization structure and reporting systems
  - Employee number, shifts, and profiles in terms of skills and experience

(b) Resource-related information
- Material procurement data and any inventories maintained
- Consumption patterns of utilities like water, steam, fuel, and electricity (This data may be collected on a monthly averaged and daily averaged basis)
- Supplier information/quality of raw material

(c) Production-related information (for industries)
- Description of technologies and processes used
- Equipment details, utilization, and conversion efficiencies
- Quality control related information in terms of quality thresholds and percentage of rejects
- Productivity-related norms and benchmarks

(d) Service-related information (for hotels)
- Services provided
- Services utilized
- Levels of services

(e) Material- and product-related information
- Packaging and disposal
- Supply chain requirements

(f) Environment-related information
- Waste collection/conveyance system
- Waste treatment and disposal operations
- Any common waste collection, treatment, and disposal systems in the neighborhood
- Any environmental measurements of the working environment (e.g. noise, air pollutants, odors)
- Any records on occupational safety and health
- Any environmental data on quality of air, water, and land in the neighborhood
- Applicable environmental regulations
- Local government policies
- Any relevant national agreements made on international issues

(g) Financial information
- Balance sheets and income–expenditure cash flows
- General cost structure showing contributions of labor, water, steam, fuel, electricity, and raw materials
- Unit costs of labor, water, steam, fuel, electricity, and raw materials
- Capital and operating costs of waste collection/conveyance, treatment, and disposal systems

One-to-one meetings aid data collection and are a more convenient method. Documented information such as operating manuals may be different to actual operational practices. By having one-to-one meetings such discrepancies can be avoided. The required questions should be asked by the right people so as to ensure that the answers obtained are valid and reliable. Many issues that arise during the review of documents and the walk-through survey can be clarified through discussions with concerned staff members (APO, 2004).

In order to determine the economic losses that result due to the generation of wastes, it is useful to identify the waste streams in the walk-through survey and to assign costs to them. Such a strategy helps in prioritizing which waste streams need to be tackled first and what types of options need to be looked at to obtain cost reduction.

The activities of gathering, analyzing, and updating data as outlined in this step are made even easier through the use of appropriate GP tools and techniques such as eco-mapping, brainstorming, attribute analyses, responsibility matrices, checklists, flowcharts, flow diagrams, material balances, and benchmarking.

2.2.2 Step 2: Planning

Planning consists of applying the information gathered in the first step to analyze the potential for improvement by identifying problems and causes. Objectives and targets can be established in order to help guide the improvement process and to allow for the management and measurement of progress (Johannson, 2005).

2.2.2.1 Task 3: Identification of Problems and Causes

Analyzing the collected data and observations of the walk-through survey permits the identification of problems and causes. The problems may be related to aspects such as processes, inputs, waste generation, product quality, or market demand. Some common problems that may be identified in the context of GP are as follows:

- Poor product quality
- Poor equipment efficiency
- Poor capacity utilization
- Environmentally unfriendly practices
- Regulatory noncompliance

In problem identification, prevailing conditions and process parameters are compared with industry standard practices and benchmarks. In this way, deviations are identified and problematic areas are verified.

Problem identification is mainly based on:

- Process flow diagram (PFD), layout, and eco-map
- Material/energy balance
- Cost of waste streams

An eco-map can be used as a tool in identifying problem areas through a macro-level assessment that considers the plant as a whole. Eco-maps are useful in identifying various focus areas in the plant, including water, solid waste, and energy (Chan, 2009).

Benchmarking is an effective tool that allows for the comparison of processes with the standards and norms of the plant and unit operation level. PFD and material balance tools can be used together with benchmarking to identify problems at the level of plant, department, and individual unit operations.

After identifying existing problems, the next step is to locate the causes. Problem-specific cause identification can be conducted by performing a cause–effect analysis using tools such as brainstorming, Ishikawa diagrams, etc.

A problem may be instigated by several causes, just as many problems may result from a single cause. The most critical causes, or root causes, should be identified.

Identified problems then need to be prioritized in order to set objectives and targets. Criteria such as the severity of the problem, its frequency, its cost implications etc. are considered when prioritizing action. An example of prioritizing problems is shown in Table 2.2.

2.2.2.2 Task 4: Setting Objectives and Targets

Once problems and likely causes have been identified, it is then necessary to set objectives and targets (APO, 2004). Problems identified in earlier tasks need to be prioritized to set these objectives and targets. The primary criteria for prioritization could be:

- Severity – this refers to the scale and implications of the problem
- Frequency – this indicates the recurrence of the problem. An insignificant problem that repeats several times in a production cycle can cause greater damage than a severe problem once in five years
- Cost implications – costing information on the possible solution is needed so as to check affordability before setting an objective
- Estimated cost of inaction – lastly, but most importantly, this is the cost of not taking any action on a problem and is estimated by costing the waste streams

A final score based on the summation of the above parameters would compare the problems on a macro-level and aid in deciding the necessary objectives and

TABLE 2.2
Examples of Problem Prioritization

Problem	Severity	Frequency	Cost of resolving problem	Cost of inaction (waste stream costing)	Score (1–10)
		Productivity			
Material consumption	High	Always	High	High	9
Product quality	Medium	82% success	Moderate	Moderate to high	6
Housekeeping	High	Always	Minimum	Moderate	10
Legal compliance	Medium to high	Occasionally on air, always on effluents	High	High	7
		Business			
Numerous rejects from customers	Minimum	5%	Minimum	High	4

targets. These need to be specific, measurable, attainable, relevant, and trackable (SMART).

The following points should be considered in setting objectives and targets:

- Objectives should be based on identified problems
- One objective can have multiple targets that are introduced over time
- Targets should be developed based on need. For example, if legal compliance is to be sought within one year, then the target for an objective that addresses a compliance parameter should be set for one year
- Indicators to track the target should be established

The progress and performances of the targets need to be monitored and reported on. Using GP tools such as control charts and multi-variate charts, the progress can be graphically indicated over a particular time period.

2.2.3 Step 3: Generation, Evaluation, and Prioritization of GP Options

GP options should be generated and evaluated by considering the objectives and targets that are formulated in the second step. GP options are generated through the integration of GP tools and techniques (Johannson, 2002).

2.2.3.1 Task 5: Generation of GP Options

The appropriate sequence of actions to be followed in generating and selecting suitable GP options is as follows:

(i) Generation
(ii) Screening
(iii) Evaluation
(iv) Final selection

Reviewing existing options is an important step for generating new GP options: it facilitates the evaluation of what has already been implemented in the organization. A more detailed review is required to identify options that are not yet implemented. The review should inquire as to the reasons for not implementing those options previously in order to evaluate their potential for implementation going forward. The options that did not succeed at all, or did not perform as expected, could be reviewed and improved by undertaking a causal analysis (APO, 2004).

By using brainstorming as a tool to analyze problems, causes, and any available information, new GP options can be generated. Brainstorming sessions should be allowed to proceed freely and without interfering the communication of new and creative ideas and options.

2.2.3.2 Task 6: Screening, Evaluation, and Prioritization of GP Options

As the next task, an initial screening of generated GP options is undertaken to consider the feasibility of the options in terms of economy, technology, environment, and safety (Parasnis, 2003).

Screening is undertaken with consideration of the following criteria:

- Cost of implementing and maintaining the option and the cost of inaction
- Technical feasibility/complexity – this relates to the technical expertise required within the industry to implement the option
- Risk – this refers to the risk of failure as compared to the investments, expected benefits, or physical hazards that may result due to implementation
- Time required for implementing – this is to establish a link between the time necessary for implementation and realization of the option, and the target set for the objective of that option
- Benefits – it is important to determine the option's potential for resolving the problem and meeting the targets set. Benefits can be broadly divided into economic, environmental, and social

Screening allows for the elimination of unsuitable or conflicting GP options as well as the prioritization of options based on certain criteria. This minimizes the wasted effort of conducting detailed evaluations of unsuitable GP options.

The Sieve Method can be used for first-level screening of GP options. Basic screening criteria such as cost, time, and complexity cut-off values are set, taking into account their affordability for the organization. GP options are compared with cut-off values and the options that are beyond those values are rejected.

Screening facilitates the elimination of GP options that:

- Are very expensive or demand considerable resources that are not presently available
- Are complicated and demand special skills that are not available
- Need to be deferred despite having a good potential, because enough information does not exist
- Pose a risk to the production process and product quality
- Are not proven and, hence, may require a phased implementation, such as lab-scale trials or pilots

Once the screening is completed and in order to evaluate the selected options, substantial option-specific information may be required (technology details, operating guidelines, fabrication and design details, cost information, etc.). Specific information regarding the nature of screened options may be collected by sub-teams, as shown in Table 2.3.

Based on the collected information, an evaluation of options should be conducted using the various applicable tools. The cost–benefit analysis is one of the most significant tools required in the evaluation of options. A cumulative evaluation of options should be conducted to ensure that options do not conflict or minimize each other's benefits.

Technical affordability is evaluated considering following aspects:

- Nature of requirements to implement the option (space requirements, utilities, and operators)
- Feasibility of technology
- Necessary process modification

TABLE 2.3
Example of Information Collection

GP Options	Option-Specific Information Required
New equipment	Suppliers, installation space, cost (capital, operating), people power requirement
New material/chemical	Quantity, cost, effects on other processes, safety sheets, transport, suppliers
Change of supplier	Reliability, costs, effect on market, other clients of the new supplier
Modification in existing process	Installation, fabrication, commissioning requirements, costs
Change in operating practices	Effects of new practice, changes due to new practice, time scheduling

Environmental soundness is evaluated on the basis of:

- How much environmental improvement/waste reduction is expected from the option
- The nature of the benefit (improved productivity, odor control, better health and safety)
- Whether the solution is short term or long term (will it be easy to adopt to new regulations?)

Financial viability is evaluated on the basis of:

- Investment requirements
- Financial viability
- The source of investment and difficulty involved in acquiring it

2.2.4 Step 4: Implementation of GP Options

The implementation of green productivity options involves two actions: preparation and execution. Preparatory work includes awareness building, training, and the development of competencies. If new equipment or systems are required, these are installed, and operator instruction and training is provided to ensure success (Johannson, 2002).

2.2.4.1 Task 7: Formulation of GP Implementation Plan

A GP implementation plan should be formulated by considering the following components:

- Nature of the option in relation to objectives and targets
- Actions needed for implementation
- Responsible person or department
- Timing and means of implementation

Therefore, during the process of originating a GP implementation plan, it is necessary to collect relevant background information for each GP option regarding:

- Relevant tools and techniques integrated in GP options and their nature (housekeeping, recycle, reuse, recovery, process/equipment modification, change in raw materials, end-of-pipe treatment, etc.)
- Points and locations for applying the option
- Option prerequisites (any linkages to success of other options)
- Required resources and means of procurement (materials, equipment, information, expertise, and finance)
- Timing and means of implementation (first lab-scale, then pilot or direct full-scale implementation)
- Risk mitigation plans and insurance measures
- Necessary support personnel and any procedural requirements to that effect
- Responsibility matrix and task allocation
- Monitoring program and the setting up of indicators (a background or baseline must be recorded before implementation of the option)
- Milestones to be achieved in the implementation sequence

Thus, GP implementation plans fulfill three major purposes, including guidelines for the implementation of selected GP options, means for management to review the option performance, and means of training and awareness building. Gantt charts are useful supportive tools during this step. An example of an implementation plan is shown in Table 2.4.

2.2.4.2 Task 8: Implementation of Selected Option

On some occasions, the implementation of options should be conducted on a trial and small-scale basis to reduce the impact on the existing system (APO, 2004).

During the implementation process the GP team may face difficulties such as:

- Inadequate follow-up on actions by various parties
- Poor accountability
- Lack of resources (e.g. people power, funds, and time)
- Lack of support from management
- Increase in production time causing insufficient time being allocated for implementation

In order to avoid these issues regular meetings and troubleshooting sessions are required among the parties involved in implementation. Through these meetings and discussions, the review and modification of implemented options is facilitated. In management reviews, achievements can be clearly depicted through written information or photographs that show the before-and-after effect of the implemented options.

2.2.4.3 Task 9: Training, Awareness Building, and Developing Competence

Training and awareness building are essential tasks for sustaining GP options in all levels of an organization. Each person who is involved in the operation, including

TABLE 2.4
Examples for Setting Objectives in Green Productivity

Objectives	Targets	Program	Actions	Responsibility	Time
Minimize water use	Reduce water consumption by 15% of present level within 1 year	1. Water reuse	1. Install equipment to recycle water	Tech M. Purch. M.	Jan 1 Mar 1
			2. Reuse rinse water of A in B	Tech M. Q.C.M.	Mar 1 May 1
		2. Change processes to reduce water consumption	1. Recycle cooling water	Tech M.	May 1 July 1
			2. Improve washing methodology	Q.C.M. Tech M.	July 1 Sep 1
			3. Train workers	Personnel Q.C.M. Tech M.	Jan 1 Mar 1
		3. Introduction of new machine	1. Introduction of spray gun	Tech M. Purch M.	Sep 1 Nov 1
			2. Change washing machine	Tech M. Purch M.	May 1 June 1
Minimize energy use	Reduce fuel for boiler by 10%	1. Heat recovery from hot wastewater	1. Install recovery system for make up water	Tech M.	Jan 1 Mar 1
			2. Use cooling water for cooking	Q.C.M. Tech M.	Mar 1 May 1
		2. Change process of disinfection	1. Change open heating to pressurized heating	Q.C.M. Tech M.	May 1 July 1
			2. Change temperature and time	Q.C.M. Tech M.	July 1 Aug 1

external suppliers and customers, should be guided through the training and awareness programs.

Training and awareness building can:

- Upgrade the skills of the people implementing the GP options
- Improve the competence of the personnel involved in monitoring and evaluating the GP program
- Motivate the workforce
- Build awareness about the GP program in the organization and among suppliers and customers
- Ensure there is an understanding of roles and expectations
- Demonstrate management commitment
- Monitor performance
- Identify potential system improvements

Training and awareness building programs should be conducted with appropriate planning and scheduling, to ensure they follow the proper sequence of tasks as follows:

- Assessing training needs
- Selecting suitable programs, methods, and material
- Preparing a training plan (who, what, when, where, and how)
- Implementing training programs
- Tracking and recording training programs
- Evaluating training effectiveness
- Improving training programs as needed

Training, awareness building, and competence development programs can be designed in various ways. Seminars and lectures on related topics and subject areas can be carried out. Such programs are categorized as off-the-job training sessions, while on-the-job training sessions can be conducted by experts in the sector, including managers and supervisors. Videos and posters can be used as aids to inform and motivate the employees during the implementation of GP. Operating manuals, work instructions, implementation manuals, and guidance documents can reach various levels of the organization, boosting awareness and improving knowledge of GP operations. Field visits can be organized to sites where GP has been successfully implemented. Such visits allow for the collection of information on various aspects, including: how to integrate GP practices into an organization, how to adjust infrastructure and organization structure for the implementation of GP, how to continue improvement of GP practices, how to identify problems and difficulties arising during the implementation process, and, finally, how to overcome them.

2.2.5 Step 5: Monitoring and Review

2.2.5.1 Task 10: Monitoring and Evaluation of Results

Once the selected GP options have been implemented, their monitoring and review is vital to check whether they are producing or exceeding the desired results. This includes monitoring the whole system to ensure that it is on track and performing as designed. Findings are communicated to management through reports so that corrective action can be taken accordingly (Johannson, 2002). These reports should be comprised of:

- Results and observations of monitoring and evaluation
- A comparison of post-implementation performance indicators with the expected targets and timelines
- Identification of options that have not been completed or have failed to achieve particular targets
- Cause–effect analyses for identified problems
- Corrective actions already taken

It is important to identify causes for typical problems in order to determine corrective actions. Poor communication and inadequate training are common causes of many problems.

Faulty or missing procedures, or equipment malfunction, may lead to a collapse in the GP implementation plan. Proper maintenance of equipment and processes is required to ensure that GP options achieve their targets and objectives. A lack of an appropriate and accurate implementation plan may lead to failures during the implementation stage of GP options. Unrealistic targets and inadequate preparation also result in many issues throughout the implementation procedure. The major root cause for such issues may be inadequate baseline data.

2.2.5.2 Task 11: Management Review

In response to the observations and submitted reports of the initial monitoring and evaluation processes, a management review is carried out. The overall GP methodology is reviewed by mangers and experts in order to determine whether it is progressing in the right direction, and whether targets are being achieved as per the implementation plan.

Management review is undertaken to evaluate the following aspects:

- Effectiveness of the GP options implemented
- Tangible and intangible benefits
- Financial savings achieved
- Difficulties faced when applying the GP methodology
- Areas for future improvement

The review team should be comprised of people who make decisions and who have the right information.

The management review should judge the influence of changing internal and external circumstances based on the success of the implemented GP options. Changing circumstances that are internal to the organization include new facilities, changes in products or services, new customers, etc. External circumstances include new laws, new scientific information, changes in adjacent land use, etc. These factors may act as both opportunities for, and threats to, the implementation of GP. The effectiveness, benefits, environmental performances, and overall adequacy of the GP program are considered under a detailed management review.

The frequency of management reviews should be decided on the basis of best suitability for the organization. Management reviews are the key to continuous improvement and to ensuring that GP continues to meet the objectives and targets previously set.

2.2.6 Step 6: Sustaining GP

2.2.6.1 Task 12: Incorporation of Changes into Organizational System of Management

In the process of sustaining GP in an organization, the necessity of corrective action is emphasized where required, or where it may build on existing success. For ongoing improvement of the options it is necessary to incorporate them within management strategies. The evaluation and monitoring activities of the previous step mark the foundation for identifying possible corrective actions, which mangers are acknowledged of through post-implementation reports (Parasnis, 2003). A corrective action could be:

- An action that directly intervenes in the operation of an option (the operating procedure is modified, or equipment is changed)
- A modification of the target, if it is found that the set target cannot be met with the present GP option
- A change in the objective itself. This is a rare case, as it would mean that the setting of objectives and targets was not conducted properly
- The modification of an existing concern or the generation of a new one
- A modification in the team structure and responsibilities

According to the corrective actions to be taken, existing GP documents are updated. These documentations may include operating manuals, targets for objectives, responsibility allocation of the staff concerned, training needs, and activities.

2.2.6.2 Task 13: Identification of New/Additional Problem Areas for Continuous Improvement

Due to internal and external factors, new problems can arise that challenge the GP program (APO, 2004). These factors should be identified in order to sustain the GP program within the organization.

Some common factors are:

- Price changes and shortages of resources
- Market fluctuations and lost markets
- Formation of new legislation related to the environment, products, labor, and packaging
- Improvement in operating norms and benchmarks
- New competition and the formulation of newer products
- Financial issues and changes in the cash flow of the company

Feedback is essential to keep progress on track and to respond to changing circumstances.

Observations and findings collected during monitoring and evaluation activities allow for the selection of corrective actions and the subsequent looping back to previous stages of the GP methodology for different GP options. It is not necessary to follow the entire methodology from the beginning. For example, it is possible that after the management review, it emerges that only objectives and targets have to be modified. In this case, the relevant GP team goes back to the second step, i.e. the planning step, rather than to the first. Thus, the methodology's principle of continuous improvement is made possible through feedback and through corrective/preventive actions geared towards improvement.

TABLE 2.5
GP Tools and Techniques Used in Each Step

Step	Tools and Techniques
Step 1	Brainstorming, attribute analyses, needs analyses, responsibility matrices, checklists, flowcharts, flow diagrams, material balances, benchmarking
Step 2	Brainstorming, eco-mapping, benchmarking, cost benefit analyses, fishbone diagrams
Step 3	Brainstorming, cost–benefit analyses, eco-mapping, failure mode and effect analyses, pareto charts, problem evaluation reviews
Step 4	Training needs analyses, team building, responsibility matrices, critical path method, Gantt charts, spider diagrams
Step 5	Eco-mapping, failure mode and effect analyses, charts (control tally, etc.)
Step 6	All tools are used here since the activities are looped back to the previous steps to provide consistency and to encourage continuous improvement. This empowers the people involved to build on their new knowledge with confidence of success

Numerous tools and techniques are used in each step to successfully complete the GP process. It is important to utilize the most suitable GP tools and techniques in each step of the process, as shown in Table 2.5.

REFERENCES

APO. "Designing green productivity." *Tech Monitor*, 2004, July 2004: 69–71.

Chan, K. "Productivity methodologies, tools, and techniques." *APO News*, March, 2009: 4–7.

Johannson, L. *Green Productivity*. World summit of sustainable development, Manila, Phillippines: APO, 2002.

Johannson, L. *Greening on the Go*. Guiding report, Tokyo: APO, 2005.

Gandhi, N.M.D. "Green productivity indexing: A practical step towards integrating environmental protection into corporate performance." *International Journal of Productivity and Performance Management* 55(7), 2013: 594–606.

Parasnis, M. "Green productivity in Asia and Pasific region." *International Energy Journal* 4(1), 2003: 53–61.

Suder, A. "Green productivity and management." In *Technology Management for the Global Future PICMET Conference* 3, 250–261. Istanbul, Turkey: IEEE, 2006.

3 Green Productivity Tools and Techniques

Guttila Yugantha Jayasinghe,
Shehani Sharadha Maheepala,
and Prabuddhi Chathurika Wijekoon

3.1 GREEN PRODUCTIVITY TOOLS

The green productivity (GP) approach uses many different tools for completing the six main phases of the program. Specifically, these tools facilitate the process of identifying problems, developing GP options, and taking corrective action.

3.1.1 Brainstorming

Brainstorming is a common tool used in green productivity and cleaner production for the purpose of gathering information. In this technique, a number of ideas are generated by different stakeholders to identify and solve problems, improve the situation, prioritize events, and set new strategies (Johannson, 2005). Brainstorming groups should be formed for the session by considering all stakeholders that are involved in the situation. Brainstorming has five main steps as described below:

1. Recognize and Assemble a Team for Brainstorming

Group members can be internal or external to the organization and they are selected by considering the following indicators:

- People who are affected by the problems
- People who will be affected by the solutions
- People who have the knowledge, experiences, skills, and ability to understand the situation
- Authorized persons who have the power to take actions
- A facilitator with a knowledge of conducting brainstorming

2. Select the Right Location

A place for the brainstorming session should be selected by considering many factors. Specially, it should be a comfortable place/room for all group members and facilitate face-to-face interaction. It should be possible for generated ideas to be recorded and displayed using special mechanisms such as laptops, white boards, projectors, overheads with transparencies, etc. These mechanisms should help to share ideas and notes legibly (Al-khatib, 2012; Rizi *et al.*, 2013). The brainstorming session should be conducted without any disturbances, and the room should facilitate privacy.

3. Define the Question

After completing the first two steps, the problem should be defined for the team members by the facilitator, to allow the participants to understand the situation. This will create a platform for prioritizing the problems and generating new solutions and strategies. Group members should have a chance to ask questions (Johannson, 2005).

4. Generate Ideas

When the problems are clearly understood and situation is clarified, idea generation can be started. The free generation of ideas should be promoted throughout the brainstorming sessions in order to find out best solution. Two or three rounds can be conducted so as to address the ideas of all the members, and no idea should be criticized or judged (Al-khatib, 2012).

5. Select the Best Solution

All generated ideas are categorized by considering their approach to solving the problem. Some ideas can be modified or combined according to the group members' opinions. Ideas should be shared within the group to encourage more detailed approaches. To refine the ideas, they should be discussed within the group, alongside the by exchange of supportive or opposing opinions. Finally, one or two ideas should be selected as the best solution for the problem and these can be implemented in the organization to improve its productivity (Al-khatib, 2012; Rizi *et al.*, 2013).

3.1.1.1 The Benefits of Brainstorming

- Ideas can be generated by all stakeholders, which facilitates a more ready understanding of the problem in all directions and with clear information. From there, different solutions can be identified, and even more outlandish ideas can be turned into the best solutions for the problems
- Stakeholders will be more willing to adapt to the implementation process because they were involved in generating the solutions
- Stakeholders are motivated and this creates an interest in their work
- Brainstorming is important in cases where many issues have occurred in different areas and solutions will affect the whole production process
- Stakeholders of all levels can clearly see others' point of view for finding the best solution, and this results in benefits for the whole organization
- This technique can be used for situations where other resolution strategies have not provided the desired outcome
- Brainstorming promotes the concept of "thinking outside the box", which can lead to the formation of more innovative ideas
- Brainstorming facilitates an understanding of the whole problem, alongside an awareness of the reality of the situation
- This technique provides a platform for generating ideas without judgment or criticism

3.1.2 Flowcharts

Flowcharts graphically represent the sequence of steps that need to be carried out in order to complete an activity. They are a standardized communication tool used to describe any activity through visual aids. Flowcharts are used to break down large activities into smaller levels Johannson, 2005. They simplify the steps of the process by making them accessible for all levels of stakeholders. Standardized shapes are used to draw a flowchart, each with different meanings. Hemi-elliptical symbols are used to start and end a flowchart. The text in the hemi-elliptical symbol represents the first and last steps of the activity. A rectangle is used to express instructions or the process steps of the activity. An arrow signifies a sequence or the direction of a process. Arrow heads point to the next activity that has to be completed. Diamond shapes with two arrow heads represent an activity that has two options or alternatives. Each arrow will be label with either "YES" or "NO" to signify the options. A circle with text or a number represents the splitting of a flowchart into other flowcharts. The continuation of the chart will be signaled by the same numbered or texted circle, either on the same page or another. The symbols used in flowcharts are described in Table 3.1. Figure 3.1 shows an example of a flowchart.

3.1.2.1 Main Steps to Develop a Flowchart

1. Identify and mark the boundaries of the selected activity. The start and end points clearly should be defined
2. Identify the process steps
3. Mark the sequence of the process and identify all special characteristics of the process, and any alternatives to it
4. Identify the correct symbols for the process steps and review the process
5. Complete the flowchart using the correct symbols and steps

Flowcharts are widely used to describe many types of project. Specifically, it can be used to describe GP programs to GP teams, who can then improve the GP program and steps using flowcharts.

TABLE 3.1
The Symbols Used in Flowcharts

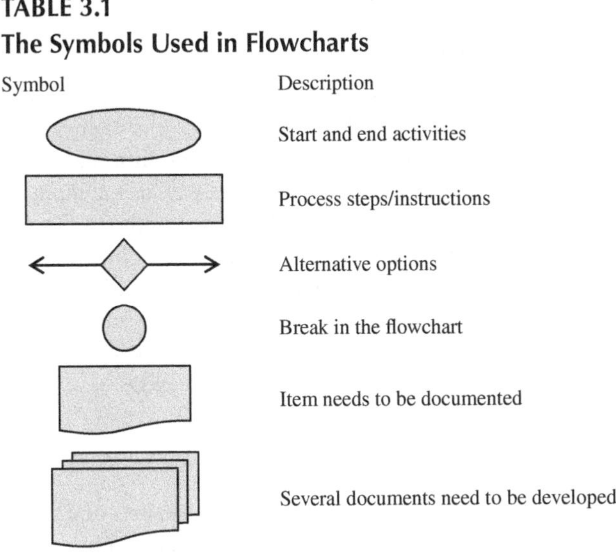

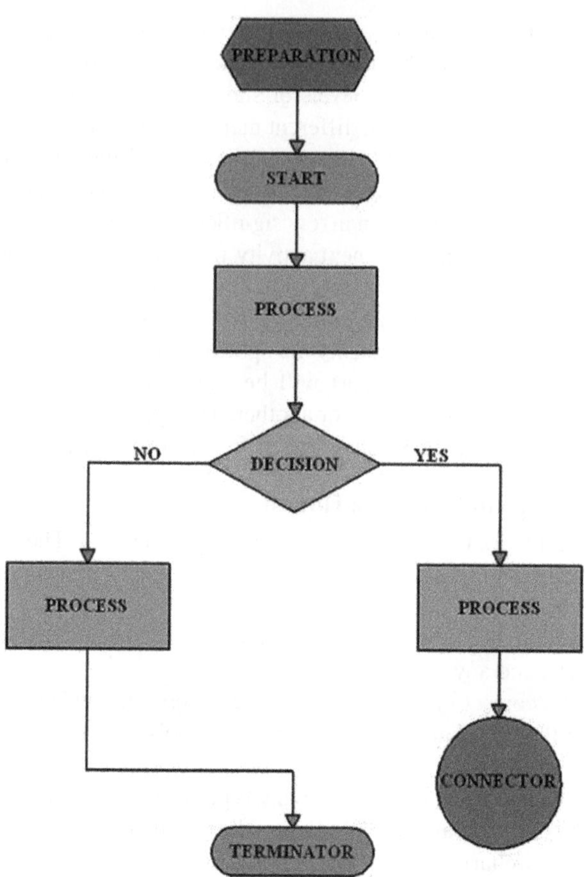

FIGURE 3.1 An example of a flowchart.

3.1.3 Process Flow Diagram

A process flow diagram (PFD) is a tool that can be used to illustrate the processes of an organization in a graphic way. It maps the mass and energy flows of an organization from external inputs to external outputs (Ibrahim and Yen, 2010). PFDs show the step-by-step sequence of the processes, as well as the inputs and outputs of each step. Energy and material amounts can be shown in the PFD to provide a more detailed picture. PFDs can clarify all details of the process and problems can be better understood by observing the diagram. PFDs clearly highlight the waste streams of an organization and can be used to process control activities (Tangkawarow and Waworuntu, 2016). Waste can be reduced and process efficiency improved by using the PFD tool in GP. Figure 3.2 shows an example of a PFD.

3.1.4 Plant Layout

A plant layout is a graphical representation of the dimensions of a building, its facilities, and their main functions. It shows the spatial relationships and activities, and

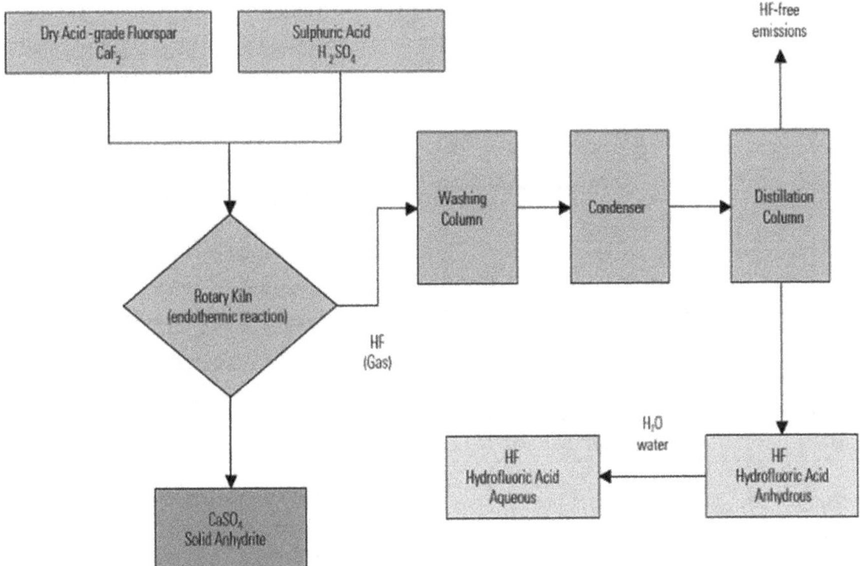

FIGURE 3.2 An example of a PFD.

their actual distribution in the building (Naik and Kallurkar, 2016). Plant layouts accurately describe a building's orientation, location, equipment, power lines, water lines, emergency ways, sewer systems, etc. The diagram should contain a title, scale, orientation, and legend to illustrate the building information as follows:

- Title – indicates the content of the image
- Scale – uses a suitable unit of measurement to represent distances
- Orientation – shows the layout orientation of the building with respect to the cardinal points, which is useful for making modifications in the building
- Legend – clearly describes all symbols used in the diagram, enabling easy identification and reading of the layout

Plant layouts facilitate planning and the making of modifications to GP programs. They are a way to detect, first hand, opportunities for improving processes and for reducing the expenses of the organization Johannson, 2005. Plant layouts provide the basis for improvements in equipment placement, leakage repairs, power line changes, and production line management. Figure 3.3 shows an example of a plant layout.

3.1.5 Eco-Maps

Eco-mapping is the visual representation of the environmental aspects of a plant. An eco-map modifies the plant layout to graphically represent situations where changes should be made to improve productivity or reduce environmental impacts (Johannson, 2005).

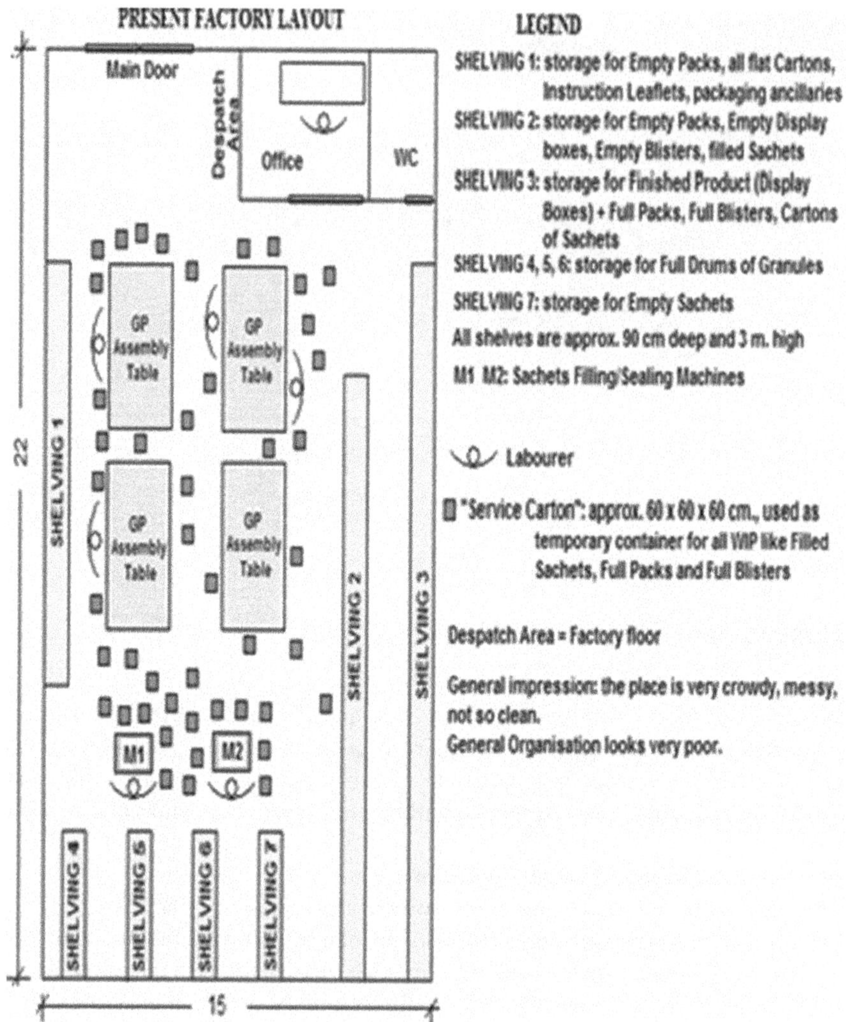

FIGURE 3.3 An example of a plant layout.

3.1.5.1 Main Steps to Develop an Eco-Map

- A plant layout is drawn to represent the whole building or a selected area. A manageable area will be used for an eco-map (e.g. one eco-map for one floor). Use one eco-map for a single issue such as water, air, solid waste, energy, etc.
- Suspected problematic areas should be addressed and marked. Actual problems (leaks, hazardous wastes, noise, etc.) must be identified, and potential problems should be marked
- Symbols can be used to represent the problems. There is no standard way to use symbols in this context, and an organization's own symbols can be used for map development

TABLE 3.2
Symbols Used for Eco-Maps

Symbol	Description	Used for
	Hatch lines	Low priority issue that should be monitored and reviewed in the future.
	A dotted circle around an area	Lower risk issue
	A solid circle around an area	A major problem. The thickness of line can be used to indicate how server a problem is
	A circle with a thick red line drawn across it	A problem that threatens the operations

Table 3.2 shows some symbols that can be used to develop eco-maps.

Eco-maps provide highly visual and readily understandable figures to highlight the problematic areas of an organization. A GP team can easily and quickly read the eco-maps and identify problematic areas to make solutions Johannson, 2005. Special techniques or training are not required to create an eco-map, which can be done within a day. The time and energy consumption in identifying a problem are very low. Eco-maps provide a clear picture for planning and selecting the appropriate solutions for an organization. Also, they can be used to monitor the process to get a clear picture of the improvements needed and carried out. Figure 3.4 shows an example of an eco-map.

3.1.6 Concentration Diagrams

A concentration diagram is a form of layout that shows areas where issues are occurring or have occurred. A GP team can focus on eliminating a particular problem by using this diagram (Bright and Newbury, 1991). Concentration diagrams can be used to investigate issues such as leaks, spills, accidents, high internal inventory, etc.

3.1.6.1 Steps to Develop a Concentration Diagram

- Begin by drafting a plant layout. A manageable area should be used to prepare this diagram
- Collect historical and actual information to develop the chart. Specifically, this information should be based on the main issues that need to be solved (e.g. energy consumption, water consumption)
- Collected information should be plotted on the layout to highlight the problems and areas where they are detected. Different icons can be used to distinguish the problems and their severity
- Areas where problems are concentrated can be identified using this diagram and these “hot areas” can be prioritized when solving problems

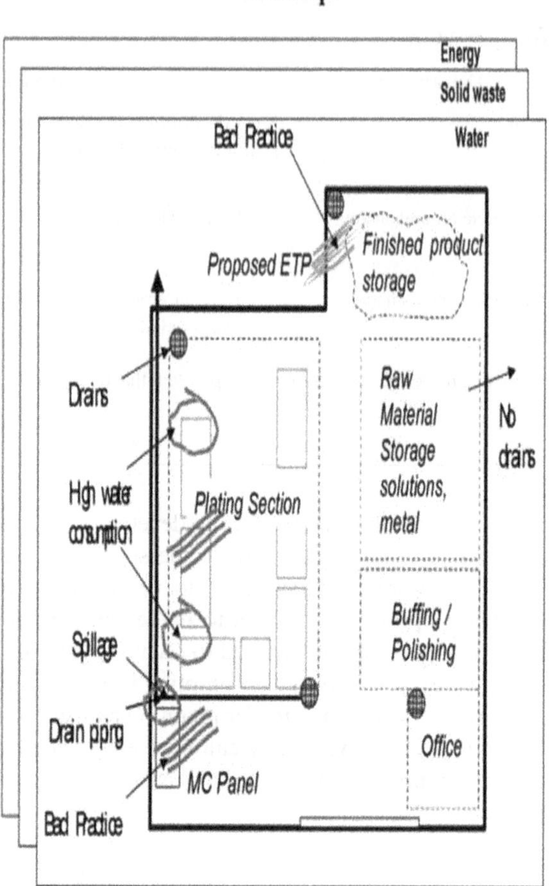

FIGURE 3.4 An example of an eco-map.

Concentration diagrams provide efficient and quick information on areas where the highest number of important issues has occurred. This can be applied to any process in an organization to give a clear idea of the weakest areas where the GP team needs to focus. Concentration diagrams give directions on good housekeeping practices to the organization to help reduce wastage and maintain sustainability. A GP team can set priorities to solve problems by observing the concentration diagram. Figure 3.5 shows an example of a concentration diagram.

3.1.7 Check Sheets

Usually check sheets are used to record information about past activities, on-going activities, or an organization's facilities, and to observe trends or patterns in this data in an objective manner (Fonseca *et al.*, 2015). They are a simple way to keep statistical data on an organization's events for the purpose of prioritizing or developing GP

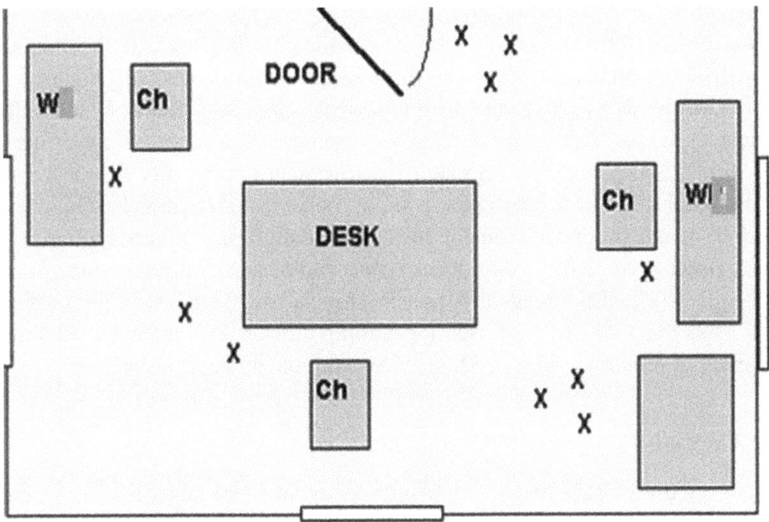

FIGURE 3.5 An example of a concentration diagram. Legend: Ch = chair, W = wall cupboard, X = problematic area.

options. Check sheets are used to observe the number of events at a given time or the number of events per location, and these can be used as inputs for concentration diagrams, eco-maps and Pareto diagrams. Check sheets can be used to detect the number of spills per month, the rate of hazardous waste generation per hour, the number of machine breakdowns per month, the number of accidents per month, etc. A check sheet provides objective evidence about the nature of the occurrences of an event.

A check sheet should include:

- Name of the GP project
- Location
- Name of persons collecting data
- Data
- Clarifications
- Date

A table format can be used to develop a check sheet. Columns can be used to record data/defects. The total recorded information should be calculated and recorded in each row and column. Commonly, check sheets are applied in the assessment of difficult or subjective events, or events with little available information, with the aim of making them objectives in the planning stage. This research focuses on the influence of company sector and size on the level of utilization of Basic and Advanced Quality Tools. The paper starts with a literature review and then presents the methodology used for the survey. Based on the responses from 202 managers of Portuguese ISO 9001:2008 Quality Management System certified organizations, statistical tests were performed. Results show, with 95% confidence level, that industry and services have a similar proportion of use of Basic and Advanced Quality Tools. Concerning size, bigger companies show a higher trend to use Advanced Quality Tools than smaller ones. For Basic Quality Tools, there was no statistical significant difference at a 95% confidence level for different company sizes. The three basic Quality tools with higher utilization were Check sheets, Flow charts and Histograms (for Services) or Control Charts/ (for Industry), however 22% of the surveyed organizations reported not using Basic Quality Tools, which highlights a major improvement opportunity for these companies. Additional studies addressing motivations, benefits and barriers for Quality Tools application should be undertaken for further validation and understanding of these results.(Fonseca *et al.*, 2015). Also, check sheets can be used for quick assessment processes to quantify events and identify patterns for statistical analysis. Check sheets are low-cost tools that can easily be used for making quick decisions and for monitoring the accuracy of processes. Figure 3.6 gives an example of a check sheet.

3.1.8 Checklist

Checklists are used for the inspection or monitoring of processes to assure all requirements have been met. This helps to reduce the variability between two inspection processes and to provide an effective, solid framework for assessment processes (Johannson, 2005). Commonly, checklists consist of a series of questions or verification points that must be monitored. Checklists should be developed for:

- Maintenance activities
- Monitoring activities
- Changing equipment
- Changing processes or operations

New GP options such as material inventories, environmental compliance, training, and human resource development can be implemented using checklists as pointers. Checklists provide a guide for the GP team to review the GP programme and permit

	Monday	Tuesday	Wednesday	Thursday	Friday	Saturday	Sunday
Wrong orders	///	𝍸	𝍸 𝍸 𝍸 𝍸 //	/	//	////	𝍸 //
Reworked orders		/	//	///		/	//
Late deliveries	𝍸 ///	/	///	//		///	//
Shipping damage						𝍸 𝍸 𝍸 𝍸	𝍸 ///
Late payments		/					
Totals	11	8	27	6	2	28	19

FIGURE 3.6 An example of a check sheet.

the easy gathering of any information needed for future analyses. An example of a check list is shown in Figure 3.7.

3.1.9 Material Balance

Material balances are developed based on the simple principle that matter is neither created or destroyed, only transformed from one state to another (Ma *et al.*, 2015). The material balance is a technique used to determine the form of materials present in the process. Materials can be categorized into three main forms: inputs, outputs, and wastes.

Inputs consist of:

- Raw materials
- Water
- Chemicals

FURNACE - 21 POINT CHECK LIST		
FURNACE		INITIALS
1	CLEAN & CHECK BURNERS	
2	CLEAN & CHECK PILOT ASSEMBLY	
3	CHECK FLAME SENSOR READING	
4	CHECK IGNITOR RESISTENCE	
5	CHECK & ADJUST GAS PRESSURE	
6	CHECK FOR PROPER TEMPERATURE RISE	
7	CHECK FURNACE FILTER	
8	CHECK & CLEAN BLOWER MOTOR	
9	CHECK DRAFT FAN AMP DRAW AND OPERATION	
10	INSPECT HEAT EXCHANGER FOR CRACKS OR CORROSION	
11	CHECK T-STAT CALIBRATION AND OPERATION	
12	CHECK T-STAT HEAT ANTICIPATION SETTING (IF APPLICABLE)	
13	LUBRICATE BLOWER MOTOR WHERE NECESSARY	
14	ADJUST PRIMARY AIR TO BURNER WHERE NECESSARY	
15	CHECK HIGH LIMIT OPERATION	
16	CHECK FAN CONTROL	
17	CHECK FOR FLUE OBSTRUCTIONS	
18	CLEAN FRESH AIR INTAKE SCREEN	
19	CHECK FOR PROPER BURNER OPERATION	
20	CHECK FOR PROPER IGNITION UPON A HEAT CALL	
21	CHECK ALL SAFTEY CONTROLS	

FIGURE 3.7 An example of a checklist.

Outputs consist of:

- Products
- By-products

Wastes consist of:

- Air emissions
- Effluent
- Solid waste
- By-products
- Chemicals

Material balances highlight the amount of missing material in a process by calculating the total materials in the process, as demonstrated in Equation (3.1):

$$\text{Inputs} = \text{Outputs} + \text{Wastes} \tag{3.1}$$

3.1.9.1 Steps to Develop a Material Balance

1. All inputs in the production process should be determined, and quantitative amounts of inputs should be recorded
2. Outputs should be identified and quantified, with calculations performed to quantify the materials
3. Waste, emissions, by-products, etc. should be taken into account
4. A tie compound should be selected
5. A preliminary material balance should be developed
6. The material balance should then be evaluated and refined

A tie compound is a measure of economic or environmental importance that can be specific to the industry (Ma *et al.*, 2015). For example, water could be selected as the tie compound in a textile industry. Material balances are a fast and economical way to identify the flows between an organization and the environment. They provide a clear picture of the inflows and outflows of an organization, as well as their quantities. They also draw attention to areas where waste may be generated in the production process. A material balance is applicable for the assessment of the environmental impact and waste generation of all processes in an organization. A material balance helps to plan for the reduction of the expenses, wastes, emissions, and environmental impacts of an organization. Figure 3.8 shows an example of a material balance.

3.1.10 Energy Balance

This is a similar concept to the material balance. Like materials, energy cannot be destroyed or created, only transformed into another form (Ma *et al.*, 2015). As a result, the total amount of energy in an organization should be a constant value, as demonstrated in Equation (3.2):

$$\text{Input energy} = \text{output energy} + \text{used energy} \tag{3.2}$$

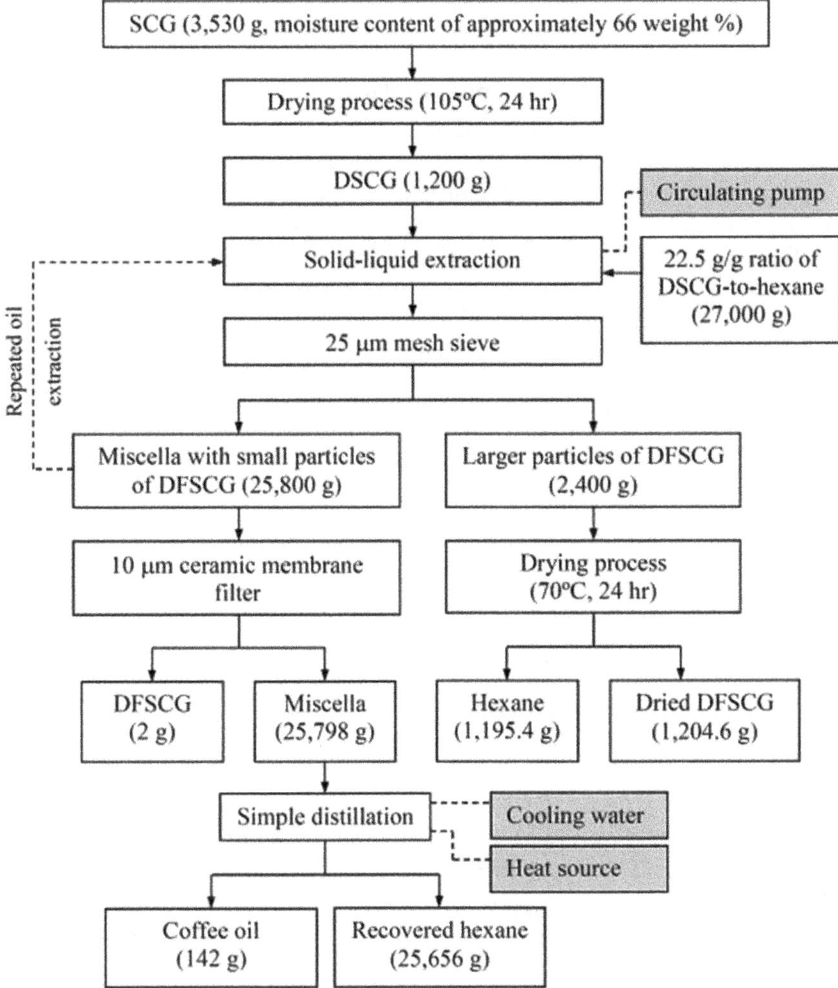

FIGURE 3.8 An example of a material balance.

Input energy can be in the form of:

- Electricity
- Fuel
- Steam

Output energy can be in the form of:

- Products
- Thermal energy
- Wastewater

An energy balance can be developed by an organization to identify areas of energy consumption and energy wastage. Energy balances provide guidance for making decisions on improving the energy usage of an organization. Using this technique, energy flows can be quantified and energy losses can be minimized in the GP process.

3.1.10.1 Steps to Calculate an Energy Balance

1. Develop a process flow diagram for the production process
2. Establish a calculation base for the process (e.g. 1000 units of product)
3. Select the appropriate energy units for the calculation process
4. Transpose all energy feedstock, needs, and usages into the selected unit
5. Calculate energy losses by subtracting the energy needs form the energy feedstock
6. Identify the areas and processes where energy losses can occur
7. Classify the losses according to their importance

Energy balances can be used to speculate the impacts of energy losses and improvements in the production process. This is useful for planning and monitoring processes in a GP program, and for maintaining organizational sustainability. Figure 3.9 shows an example of an energy balance.

3.1.11 Cause-and-Effect Diagram (Fishbone/Ishikawa diagram)

Cause-and-effect diagrams were designed by Ishikawa to determine the relationship between an event and its potential causes. The diagram permits a systematic perspective on the root causes of an issue and helps to identify the best ways to solve the

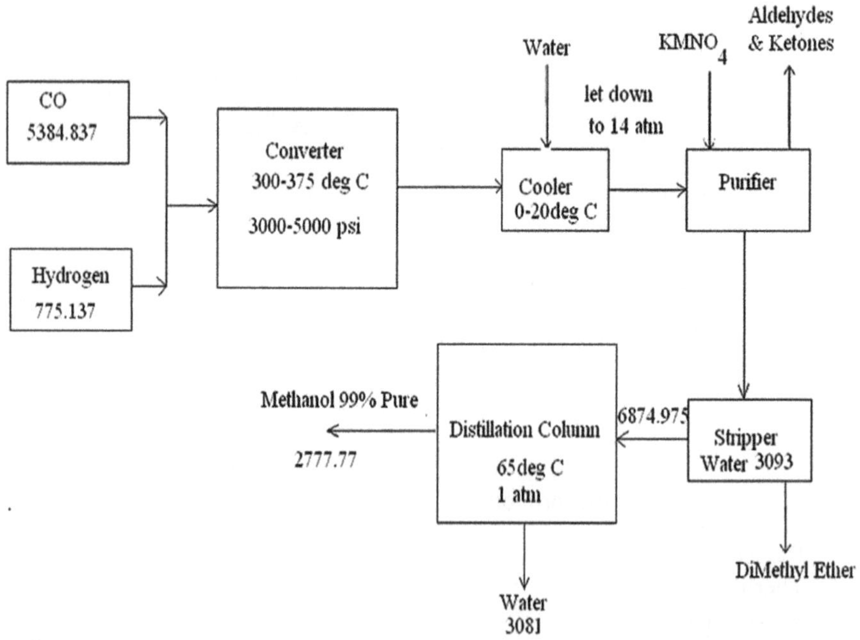

FIGURE 3.9 An example of an energy balance.

problem (Liliana, 2016). Specifically, problems and their likely causes are identified by the people who deal with them in their day-to-day lives, as they are likely to have the best solutions for them. The occupants who are familiar with the selected issue should participate in developing the diagram.

Cause-and-effect diagrams enhance the thinking process and provide all angles of a problem, leaving aside any personal viewpoints (Liliana, 2016). This enables the GP team to study all causes of an issue, decide the major and minor causes, select controllable and uncontrollable causes, and make improvements to control the causes. Cause-and-effect diagrams will thus provide GP programs with good guidance on specific targets. They can also be used for situations where multiple causes and issues are generated (Liliana, 2016). When a problem has diverse factors that may cause it, and these causes have complex relationships, the GP team can use the cause-and-effect diagram to easily understand all causes and effects of an event. This diagram is an efficient, low-cost, and low-resource-consuming tool that offers many benefits to the organization that uses it.

3.1.1.1 Steps to Develop a Cause-and-Effect Diagram

- Clearly define the issue and where it occurs
- Write the issue and location in a box placed on the center-right of the paper
- Attach a horizontal arrow to the box, with five diagonal arrows emerging from it to create fishbone-like structure as in Figure 3.10.
- Assign a label to each diagonal arrow: "man", "machine", "material", "method", and "environment" (4M-1E).
- Find all causes for the issue and categorize them into the 4M-1E structure
- Draw horizontal arrows to the diagonal arrows and write the causes of each category on them, as shown in Figure 3.11.
- Identify causes clearly and easily by using the cause-and-effect diagram

Figure 3.12 shows an example of a fishbone diagram.

3.1.12 Control Charts

A control chart is a graphical representation of a variable's behavior over a period of time. To be considered as acting under normal conditions, the variable should behave within two set limit levels called the upper and lower limit (Tennant *et al.*, 2007). If

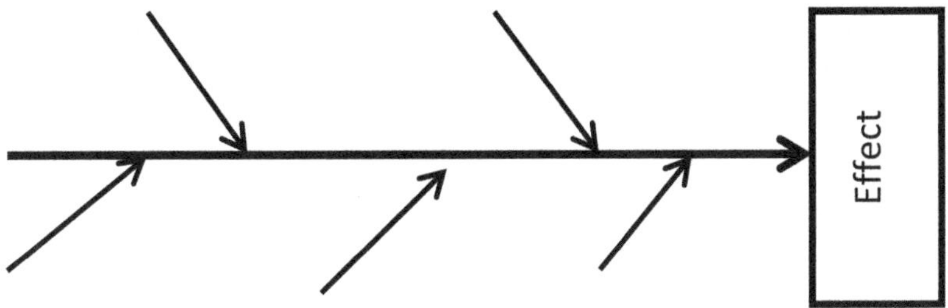

FIGURE 3.10 Diagonal arrows used to develop a fishbone-like structure.

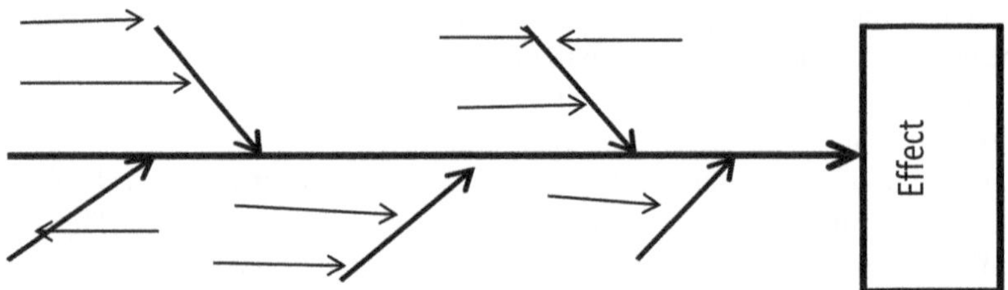

FIGURE 3.11 Horizontal arrows used to complete the cause-and-effect diagram.

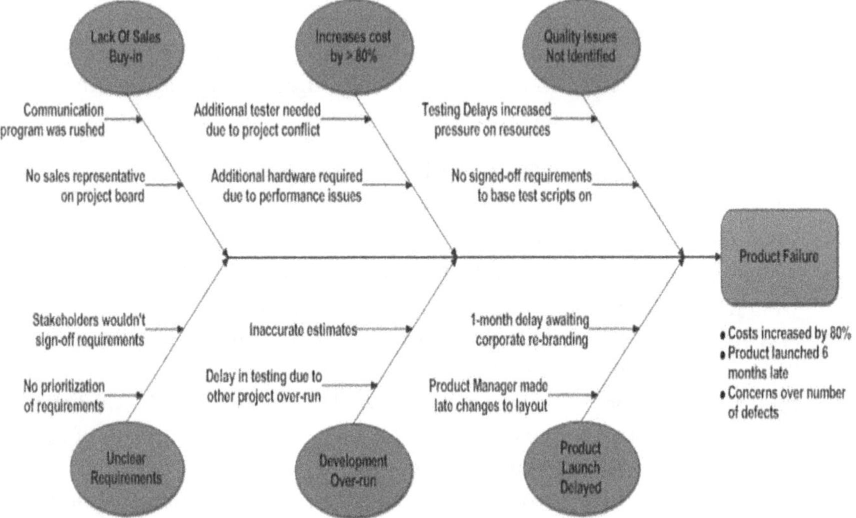

FIGURE 3.12 An example of a cause-and-effect diagram.

the variable's behavior exceeds these limits, this signals an unusual condition and some issues will likely be identified. Control charts are used to identify issues in a process and monitor the effectiveness of an improvement in an organization. Further actions or improvements may be needed to eliminate the identified issues (Tennant *et al.*, 2007). Control charts also facilitate the identification of critical parameters in a process that can lead to system failures. Thus, control charts can be considered the "pulse monitors" of an event that seek to mitigate failures. They can be applied to any parameter that needs to improve its efficiency or prevent problems from occurring.

The main three factors needed to develop a control chart are:

- A variable parameter to be measured
- A minimum acceptance value (upper limit)
- A maximum acceptance value (lower limit)

When all information has been collected, the control chart should be drawn using a Cartesian axis. Variables' values should be plotted over a time period, and upper and lower limits should be marked. The GP team can then observe the control chart and identify any issues. Figure 3.13 shows an example of a control chart.

3.1.13 Spider Web Diagram

A spider web diagram facilitate a GP team's understanding of the strengths and weaknesses of the established criteria in a GP program. Each axis of the diagram graphically represents a criterion and shows the current performance, immediate target/goal, industry average, and world-class standards of a criterion (Gomiero and Giampietro, 2005). A spider web diagram enables improved control over the performances of the diverse attributes in a GP team. It can be used to increase understanding of the current level of progress and of priorities. Also, the diagram is helpful in conjunction with information collected through eco-mapping. Spider web diagrams, also known as radar charts, can called the "report card" of a GP program. They are a simple, visual aid for sharing information with stakeholders and communicating the progress of a GP program.

3.1.13.1 Steps to Develop a Spider Web Diagram

1. Use people who are not included in the GP team to develop the spider web chart
2. Select 5–10 criteria and define them. Brainstorming or the use of an affinity diagram can be useful for determining the criteria
3. Draw the center circle of the diagram with spokes. Each spoke represent one criterion. Rank the center circle as "0" (no performance) and the end of the

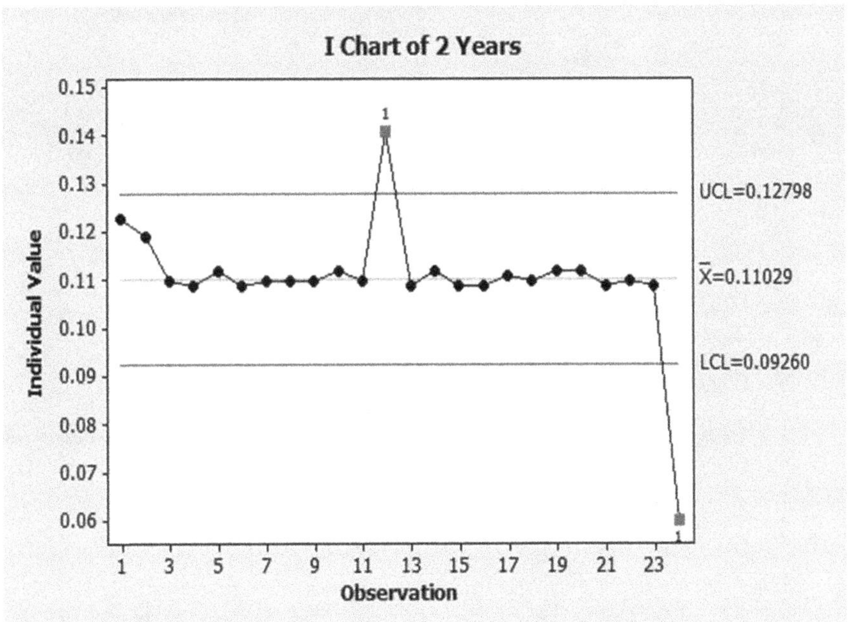

FIGURE 3.13 An example of a control chart.

spoke as the highest possible number (optimal performance). Performances can be ranked subjectively or objectively
4. Rank the criteria on the spokes as dots according to their performances
5. Connect the data and highlight the area within the connected dots. Different colors can be used to represent data from different individuals or one color can be used with larger dots
6. Discuss the results to ensure the accuracy of data. Figure 3.14 shows an example of a spider web diagram

3.1.14 Cost–Benefit Analysis

The cost–benefit analysis was introduced by a French engineer, Jules Dupuit, in the 1840s. Simply put, it is a financial tool that analyzes the total costs and benefits of an organization (Haefeli *et al.*, 2008). A cost–benefit analysis is used as a direct tool to decide whether or not to pursue a GP program. The analysis is carried out using financial values, and intangible items can be included. The analysis translates productivity and environmental benefits into mandatory units to facilitate their comparison. Pareto efficiency can be improved by attaining a low cost–benefit ratio in an organization. The values of a cost–benefit analysis are determined by the accuracy

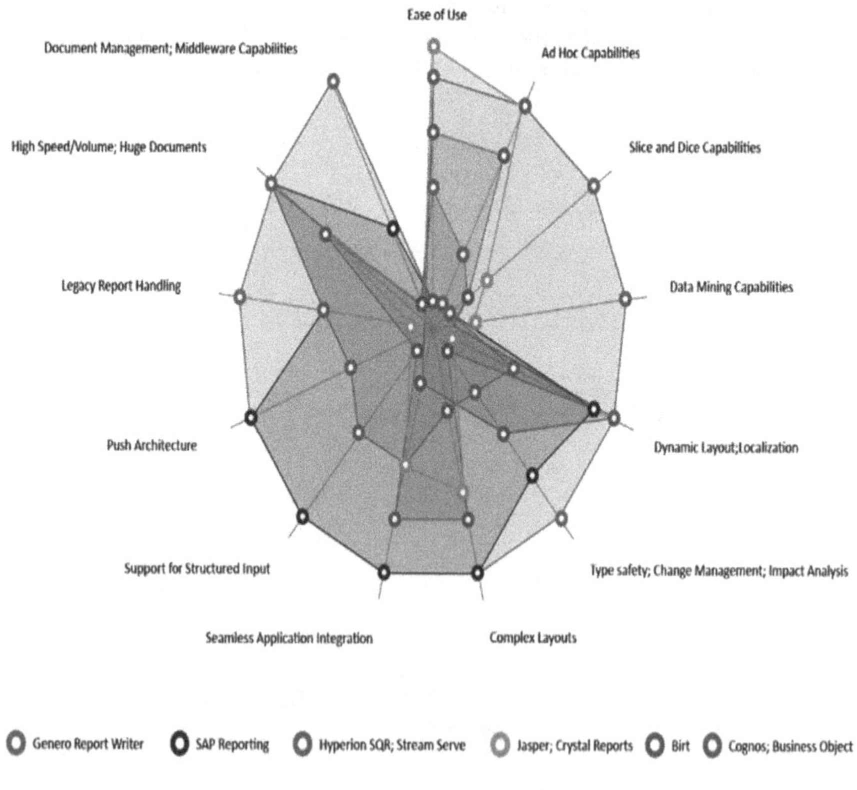

FIGURE 3.14 An example of a spider web diagram.

of the individual cost and benefit estimates. The payback period of a GP program can be identified by this analysis. To use the tool, first list all anticipated costs associated with a project, and then estimate the benefits likely to be received from it.

3.1.14.1 Steps to Produce a Cost–Benefit Analysis

1. All costs and benefits of the program are identified using brainstorming sessions
2. All costs involved in all phases of program, including the costs of transportation, purchasing (material, energy), and human effort, are analyzed. Costs are often relatively easy to estimate
3. Benefit analysis is more difficult than cost analysis because it is less straightforward to predict revenues accurately, especially for a new program. Also, financial benefits can include intangible, or "soft", benefits that are important outcomes of the project. For example, environmental impacts, employee satisfaction, or health and safety improvements can be included as benefits, but it is difficult to assign monetary values to these impacts
4. Finally, the value of costs is compared to the value of benefits. This analysis can be used to determine the suitability of a new GP program. If benefits outweigh the costs, a GP program can be implemented in the organization. At this stage it's important to consider the payback time, as calculated by Equation (3.3), to find out how long it will take to reach the breakeven point (the point in time at which the benefits have just repaid the costs)

$$\frac{\text{Total cost}}{\text{Total benifit}} = \text{Payback period} \tag{3.3}$$

3.1.15 Benchmarking

Benchmarking is a tool used to identify the best practices to maintain the sustainability of any process in an organization (De-Castro and Frazzon, 2017). Benchmarking facilitates the comparison of the processes and performance metrics of an organization with industry-best practices from other companies. The selection, planning, and implementation of a GP program is supported by the benchmarking process. Benchmarking can also be used to measure the performance of a program by using a specific indicator to create a metric of performance that is then compared to other metrics. In this process, a specific value/practice is identified as a benchmark for a process, and is then considered as the best practice (Johannson, 2005).

There are three specific types of benchmarking: process benchmarking, performance benchmarking, and strategic benchmarking. An organization should not limit its scope to its own industry when making use of benchmarking. There is no standard way to develop a benchmarking process.

3.1.15.1 Typical Benchmarking Methodology

1. Problematic areas of the organization are identified, and parameters that need to be benchmarked are decided
2. A benchmark team is formed
3. Other industries that have similar processes are listed

4. Organizations that are leaders in these processes are identified, and their best practices are observed
5. Surveys are conducted to identify best practices, specific targets, alternatives, and leading companies
6. Leading practices are identified through these studies
7. The organization's new practices are improved using the survey information. The best practice most suitable for the organization is identified and implemented

The 12-stage approach to benchmarking developed by Robert Camp is as follows:

1. Select the subject
2. Define the process
3. Identify potential partners
4. Identify data sources
5. Collect data and select partners
6. Determine the gap
7. Establish process differences
8. Target future performance
9. Communicate
10. Adjust the goal
11. Implement
12. Review and recalibrate

The benchmarking process gives comprehensive targets for identifying the success of a GP program. Also, it offers new and innovative ideas for improvements. Importantly, benchmarking is only applicable for normalized parameters. Figure 3.15 shows an example of benchmarking.

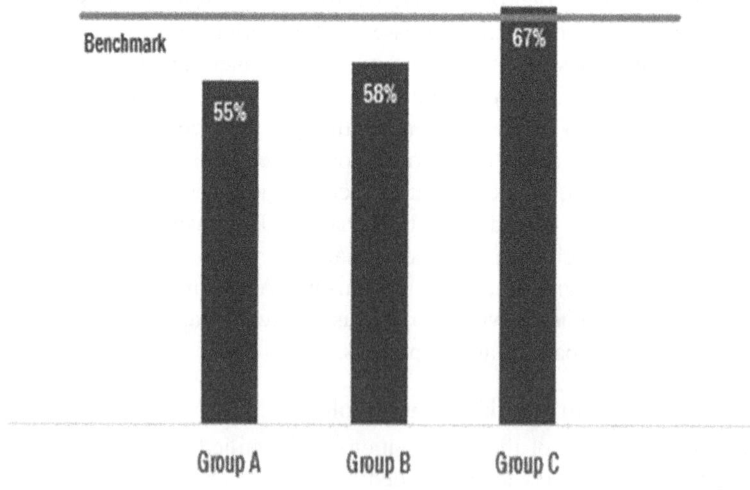

FIGURE 3.15 An example of benchmarking.

3.1.16 Decision Matrix

A decision matrix is used to convert qualitative data into quantitative data by using evaluation criteria. A decision matrix uses a number of values in rows and columns to systematically identify, analyze, and rate the performance of relationships between sets of values and information (Olabanji and Mpofu, 2014). Decisions that are based on the selected decision criteria are considered the elements of a decision matrix. Usually, this tool is used to describe a multiple-criteria decision analysis (MCDA) problem.

When a clear and obvious preferred option does not exist, the analysis of a decision matrix helps in selecting the best decisions from number of good alternatives by considering different factors.

3.1.16.1 Steps to Develop a Decision Matrix

1. All options/alternatives and factors that need to be considered should be listed. Options and factors should be written in columns and rows respectively on a table
2. A scale should be selected to weight all criteria. For example, the scale might range from 0 (poor) to 5 (very good). If none of the criteria are met for a particular factor in your decision, then all options should score 0
3. Identify the relative importance of the factors in each decision. Rank these as numbers using the scale according to their importance for the final decision (0 – unimportant, 5 – very important)
4. Next, multiply each of the weighted criteria by the values that signify the relative importance of the factor. This gives a weighted score for each option/factor combination
5. All weighted scores should be added to each of the options, and the highest scored option should be identified as the best one

A decision matrix helps to analyze several options by considering different factors. There is no standard way to design the table or scale, and thus the GP team can design their own table and scale for developing the decision matrix. Figure 3.16 shows an example of a decision matrix.

3.2 GREEN PRODUCTIVITY TECHNIQUES

GP techniques are used to make changes that result in a better environmental performance and improved productivity in an organization. They range from simple techniques (housekeeping) to complex techniques (designing "green" products). GP techniques are organized into the following categories:

- Waste management
- Training and awareness
- Waste prevention
- Resource conservation
- Product improvement

Decision Matrix Example for Battery			ENTER SCORES	Extend Old Battery Life	Buy New Batteries	Collect Experient Data With Alternative Experiment	Cancelled Experiment
CRITERIA	Mandatory (Y=1/N=0)?	Weight	SCALE				
Mission Success (Get Experiment Data)	1	30	3 = Most Supportive 1 = Least Supportive	2	3	3	0
Cost per Option	0	10	3 = Least Expensive 1 = Most Expensive	1	2	3	1
Risk (Overall Option Risk)	0	15	3 = Least Risk 1 = Most Risk	2	1	2	3
Schedule	0	10	3 = Shortest Schedule 1 = Longest Schedule	3	2	1	3
Safety	1	15	3 = Most Safe 1 = Least Safe	2	1	2	3
Uninterrupted Data Collection	0	20	3 = Most Supportive 1 = Least Supportive	3	1	2	1
WEIGHTED TOTALS in %		100%	3	73%	60%	77%	0%

FIGURE 3.16 An example of a decision matrix.

3.2.1 Improved Operating Procedures

Operating procedures help to specify processes, schedule operations, maintain procedures, and handle material in a sustainable manner. They provide safety considerations for staff and compare resource and energy values with optimum conditions (Lymperopoulos *et al.*, 2015; Freitas *et al.*, 2016).

Operating procedures can be improved by using the following steps:

1. Clearly state the purpose of procedures –describe the compliance practices, laws, regulations, and specific actions of the procedures
2. Develop a process flow diagram – clearly show the process steps in a readable way. The process flow diagram should be developed to represent a compliant and operationally efficient sequence of activities
3. Seek knowledge, experience, and willingness from functional, technical, and compliance experts to improve efficiency
4. Clearly write the procedure in a straightforward and easy to follow way. A lack of detail in procedures can cause communication problems and an inconsistency in processing
5. Provide a style guide, which is useful for authors and helps procedures to function together and serve a common purpose
6. Link processes and procedures; it is adequate for processes to be fluid across procedures
7. Clearly identify procedure (system) owners and ownership behaviors
8. Write procedures that facilitate human works and offer a comfortable means of following them

9. Conduct training in the organization to bring about change either through voluntary quality system improvement, or through involuntary regulatory enforcement action
10. Manage the change
11. Improve quality
12. Support continuous improvement
13. Increase capacity for creative work

3.2.2 Waste Segregation

Waste segregation is the process by which waste is separated into different elements. Waste segregation can be practiced in the place where waste is generated (on-point) or at the initial stage of the waste treatment process (Malik *et al.*, 2015). On-point sorting is more efficient, although waste sorting can occur manually or automatically. Waste segregation divides wastes into different elements with respect to their characteristics, such as dry/wet, biodegradable/non-biodegradable, organic/inorganic, etc.:

- Dry waste (e.g. wood and related products, metals and glass)
- Wet waste (e.g. organic waste)
- Biodegradable waste (e.g. vegetables, fruits, flowers, leaves from gardens, wood shavings, pencil shavings)
- Non-biodegradable waste (e.g. plastics, paper, glass, metal, aluminum foil, and other tetra packs)

Waste segregation facilitates efficient waste treatment and management practices. Separated waste can be treated easily using effective techniques (Malik *et al.*, 2015). The process reduces open dumping or waste accumulation in landfills. Sorted wastes can be in the form of:

- Paper
- Cardboard (including packaging)
- Glass (clear, tinted)
- Plastics
- Textiles
- Wood, leather, rubber
- Scrap metal
- Compost
- Special/hazardous waste
- Residual waste
- Organic waste

Waste segregation protects health and reduces environmental pollution. It also reduces the burden of the end-of-pipe treatment of a non-compatible pollutant stream. Also, it increases the possibility for recycling and reusing of wastes, and maintains sustainability in an organization.

3.2.3 Good Housekeeping Practices

A good working environment is important to maintain efficiency and productivity in an organization. Good housekeeping practices provide a comfortable and safe working environment for occupants. Specifically, it eliminates the inefficiencies and accident hazards caused by unfavorable conditions in the workplace (Agwu and Samuel, 2014). Good housekeeping practices should feature in every phase of industrial operations, and on the entire premises, both indoors and outdoors. Good housekeeping is not only about cleanliness but is also about maintaining orderly conditions, preventing congestion, providing adequate storage arrangements, marking passages, and ensuring a suitable establishment for maintenance.

Bad housekeeping practices can cause accidents such as:

- Slipping on greasy, wet, or dirty surfaces
- Falling over loose objects on floors, stairs, and platforms
- Equipment dropping from above
- Hitting against poorly stocked or misplaced material
- Scratching body parts on nails, wire, steel strapping on bales or crates, etc.

Some good housekeeping practices are:

- Space: the available space should be sufficient to work in for every individual and for the equipment
- Aisles: mark walkways and storage areas to reduce traffic
- Storage: acceptable and suitable space for materials and equipment should be provided. Neatness and order are essential in overcoming storage problems
- Materials handling: layouts should be planned for material flows, with efficient methods and equipment
- Ventilation: good ventilation should be provided to remove air contaminants at the source using exhaust air
- Lighting: lighting should be well distributed and efficiently operated. Dirty lamps and shades should be cleaned, and essential light fixtures for employees should be provided. Windows should be kept in clean to let light in
- Floors and walls: use materials that are easy to keep clean and in good repair. Floor areas should be kept clean by removing spills, chips, dust, and wastes to reduce accidents. Walls should be painted using suitable colors to boost the environment
- Amenities: clean, up-to-date washrooms and lockers for clothing should be provided
- Waste removal: practice waste removal to prevent congestion and disorder
- Scraps and spillage: scraps should be disposed of and spills should be prevented or cleaned. Prevent accidents by keeping oil and grease off the floor
- Dust and dirt: In some jobs, dust, dirt, chips, etc., are unavoidable. Vacuum cleaners can be used to clean them
- Meal rooms and restrooms: maintain a high standard and clean environment in meal rooms, restrooms, etc.

- Tools and equipment: tool housekeeping is very important. Tools should be placed in an orderly and neat fashion either in the tool room, on the rack, or out in the yard using suitable fixtures. A regular system of inspecting, cleaning, and repairing tools is an essential part of any program
- First aid: first aid facilities and equipment should be kept under perfectly clean conditions and be fully stocked
- Fire-control equipment: fire-fighting equipment (extinguishers and fire hoses) should be frequently inspected and kept in good working condition
- Maintenance practices: these are highly important for good housekeeping
- Assign responsibility: employees should be assigned with the responsibilities of good housekeeping practices, such as cleaning. Additional staff can be hired for the purpose of cleaning
- Checklist: prepare a checklist that suits the requirements of the workplace to maintain the good housekeeping practices

3.2.3.1 Benefits of Good Housekeeping

- Reduced number of accidents and improved worker safety
- Decreased fire hazards because of a reduction in the presence of flammable materials (waste, dust, debris)
- Reduced number of workers being struck by objects, which is a result of the organized storage of materials, tools, and equipment
- Effective use of space
- Reduced worker injuries, which is a result of the proper maintenance of machinery, equipment, or systems
- Reduced worker exposure to hazardous substances
- Improved worker health
- Increased worker productivity
- Reduced additional costs
- Better control over tools and materials, as well as over the inventory of supplies
- Improved working conditions
- Reduced property damage

3.2.4 The 5S Technique

5S is a workplace organization method that was developed based on a Japanese concept. It consists in five Japanese words: *seiri*, *seiton*, *seiso*, *seiketsu*, and *shitsuke*. 5S describes how to organize a workspace efficiently and effectively (Michalska and Szewieczek, 2007). It helps to improve the productivity and efficiency of employees and organizations. Table 3.3 describes the 5S concept.

3.2.5 The Concept of 7 Wastes

The concept of 7 wastes is known as "muda", and was developed by Toyota's Chief Engineer Taiichi Ohno. Also known as the lean manufacturing process, it focuses on

TABLE 3.3
The 5S Method

5S Element	Meaning	Actions
Seiri	Sort	Eliminate obstacles, remove waste, segregate unwanted materials and equipment from the workplace and remove them. Evaluate necessary items with respect to cost or other factors. Use red tags to mark areas where unwanted items are present but cannot immediately be disposed of. Dispose of these items when possible. Regular monitoring is needed.
Seiton	Set in order/ Simplify	Arrange all necessary items to prevent loss or waste of time. All tools/ equipment should be in close proximity to make it easy to find and pick them up. Implement a first-in-first-out (FIFO) process.
Seiso	Shine/Sweep	Clean workplace with tools and resources. Use cleaning as an opportunity for inspection, to prevent machinery and equipment deterioration
Seiketsu	Standardize	Establish procedures and schedules to ensure the consistency of implementing the first three 'S' practices. Use photos and visual controls to keep things in order. Use standardized color codes for usable items.
Shitsuke	Sustain	Perform regular audits. Enforce training and discipline.

eliminating waste in the production process. Before eliminating waste, it is important to identify the points where wastes are generated. While products and processes differ between factories, the typical wastes found are quite similar. For each waste, there is a strategy to reduce or eliminate its effect on an organization, and each strategy centers on improving overall performance and quality (El-Namrouty, 2013). The 7 wastes are classified as: overproduction, waiting, transporting, over-processing, unnecessary inventory, unnecessary motions, and defects.

3.2.5.1 Overproduction

When products are manufactured before they are actually required, this is known as overproduction. Overproduction causes unnecessary costs in the manufacturing process because it prohibits the smooth flow of materials and reduces quality and productivity. Overproduction is considered the worst form of waste because it leads to other wastes and hides the need for improvement.

3.2.5.2 Waiting

The waste of waiting occurs when goods are not moving or being processed. Operators and equipment in one step of the process are halted while waiting for the second step to start up. Traditional manufacturing processes meant that a product spent more than 99% of its life waiting. Poor material flow, excessively long production runs, and long distances between work centers resulted in this waste. More efficient processes can be developed to eliminate this waste and save time and money.

3.2.5.3 Transporting

Transporting products to the sites where processing takes place consumes extra time and money. Excessive movement and handling can result in damaging the materials and quality of the product. Material handlers with proper methods and equipment must be used to transport materials, and hence transportation is difficult to reduce. However, excessive movements do not directly correspond to the value-adding process. They should be minimized to reduce delays and the risk of handling, and to remove non-value-adding process steps and costs.

3.2.5.4 Inappropriate Processing

Over-processing waste adds more work/steps into producing the product, which causes more cost than the customer values. Many organizations use expensive high-precision equipment and processes where simpler tools can be used. In addition, they encourage overproduction with minimal changeovers in order to recover the high cost of this equipment. The use of appropriate techniques and processes will help to eliminate this waste.

3.2.5.5 Unnecessary Inventory

Work in progress (WIP) is a result of overproduction and waiting. An excess inventory:

- Increases lead times
- Consumes productive floor space
- Delays the identification of problems
- Inhibits communication
- Increases the cost of manufacturing
- Reduces available storage space
- Can cause the shelf-life to expire
- Consumes extra resources
- Increases damages

An excess inventory tends to hide issues on the plant floor thatmust be identified and fixed in order to improve operating performance. By achieving a seamless flow between work centers, many manufacturers have been able to improve customer service and slash inventories and their associated costs.

3.2.5.6 Unnecessary/Excessive Motion

Motion waste includes unnecessary movements (of a machine or employee) that are more complicated or difficult than necessary. Excess motion is an ergonomic waste that can be observed in instances of stretching, bending, lifting, walking, and reaching. It affects the health and safety of employees. Also, it can damage equipment and products. Excessive motions should be analyzed and redesigned to reduce negative consequences.

3.2.5.7 Defects

Defects have a direct impact on the bottom line, the quality of products, and the expenses of the organization. The waste resulting from defects includes quality errors that are invariably costly. Associated costs include quarantining and re-inspecting the inventory, rescheduling transportation, and capacity loss. In many organizations, the total cost of defects has often been a significant percentage of the total manufacturing cost. Defects can be reduced through worker involvement and continuous process improvement (CPI).

3.2.6 Preventive and Productive Maintenance (PPM)

Preventive maintenance means taking actions in advance, before problems occur, with regard to the functioning of equipment. Productive maintenance applies when the results of equipment maintenance are measured and the results are positive. Productive maintenance takes place when the savings are higher than the maintenance costs (Ahuja and Khamba, 2008). Preventive and productive maintenance, referred to as PPM, is useful for minimizing equipment downtime, enhancing productivity, and maximizing efficiency. It consists in maintaining equipment operations at peak condition levels to reduce negative impacts. Preventive maintenance is comprised of:

- Cleaning
- Lubrication
- Inspection of protective coating
- Replacement of parts and overhauls

A cost-effective maintenance system can be designed through:

- Classification and identification of equipment
- Collection of information
- Selection of maintenance policies
- Preparation of preventive maintenance program
- Preparation of corrective maintenance guidelines
- Organizing of maintenance

Total productive maintenance (TPM) is considered an advance step of PPM that focuses on continuously improving maintenance services to provide the optimum level of all manufacturing processes. TPM is a holistic approach for achieving high-quality production without breakdowns, defects, accidents, or slow runnings (Ahuja and Khamba, 2008). The implementation of a TPM program generates a mutual responsibility for equipment that encourages greater association between plant floor workers. TPM can be achieved through:

- Planned maintenance: maintenance tasks are scheduled based on predicted or measured failure rates. This decreases inventory through better control of wear-prone and failure-prone parts
- Autonomous maintenance: this ensure responsibilities for routine maintenance. Ownership of the equipment is given to the operators by increasing their knowledge, ensuring the equipment is well cleaned and lubricated, identifying emergent issues, and providing high maintenance

- Quality maintenance: errors should be prevented, and a method should be identified to detect errors in production processes. Root cause analysis can be applied to eliminate frequent sources of quality defects
- Early equipment management: new equipment can be designed using direct practical knowledge and experience to achieve high quality and faster performance levels
- Focused improvement: small groups of employees should work together proactively to achieve improvements in equipment operation. Frequent issues are identified and solved through cross-functional teams
- Safety maintenance and a healthy working environment
- Training and education: knowledge gaps should be filled to achieve TPM goals. Operators, maintenance staff, and managers should participate in training to develop skills for maintaining equipment and identifying emerging problems
- TPM in administration: TPM techniques can be applied to improve administrative functions and extend benefits

3.2.7 Recycle, Reuse, and Recovery

Waste generation results in many negative impacts on the environment. Specifically, environmental regulations prohibit the discharge of toxic pollutants into the environment and promote resource conservation through the recovery, recycling, and reuse (3R concept) of waste products (Noll *et al.*, 1986). Material recovery from waste is better than its disposal, but whether this is feasible depends on the available economic conditions and technology Johannson, 2005.

3.2.7.1 Reuse

Reuse is described as using something again and again before throwing it away. This reduces the amount of waste generated waste amount more than normal practices.

For example, glass bottles can be used again and again for a long period of time and for different purposes before being thrown away.

3.2.7.2 Recycle

The concept of recycling describes the mechanical process a product goes through to change its form and purpose and turn it into a new product. This is only recommended when reducing and reusing are not possible.

For example, plastic chairs can be recycled to make plastic baskets.

3.2.7.3 Recover

In the recovery process, wastes are converted into resources such as electricity, heat, compost, and fuel through thermal and biological reactions. Specifically, energy is recovered through this process. The following are different methods of conducting recovery:

- Direct recovery: larger waste generators use the direct recycling process as it offers stronger economic incentives than treatment or disposal processes (for example, chromic acid recovery from spending plating bath solutions and organic solvent recovery from decreasing operations) (Noll *et al.*, 1986)

- Secondary industry recovery: the waste from a given industry may contain material of value to another industry. This material can be used as an input for another industry with or without intermediate purification or enrichment, and the process is known as secondary industry recycling (For example, phenols recovery from coking wastes, sulfur recovery from stack gas cleanup, and metal and/or acid recovery from various pickle liquors)
- Resource recovery: resource recovery offers economic and environmental benefits. The market for recovery is increasing and includes centralized commercial processing, recovery facilities, and industrial waste exchanges. Each increment of hazardous waste that is recycled has value, and this reduces the need for detoxification and ultimate disposal
- Energy recovery: many hazardous wastes (particularly those containing organic matter) have a sufficient energy value to enable the recovery of that energy to be economically viable. Commonly, spent solvents or waste oils are used for steam generation. In addition, waste chlorinated solvents are used as an alternative fuel source in cement manufacture. Energy recovery may occur at the site of the generator (on-site) or at a secondary site (off-site)

3.2.8 Energy Conservation

Energy conservation seeks to reduce the consumption of energy by using it more efficiently or by reducing the amount a service uses (e.g. driving less) (O'Rielly and Jeswiet, 2014). Energy can be conserved by:

- Performing an energy audit: the total energy usage of a building can be inspected and analyzed using energy audits. This provides information about energy consumption, wastage, energy flows, etc. that can help to make improvements in energy consumption without negatively affecting output
- Designing a passive solar building: this includes window placement and thermal insulation, glazing type, thermal mass, and shading to reduce energy consumption in the building
- Lighting: CFL bulbs (with a label that saves 80% electricity) can be used instead of ordinary bulbs. Windows should be installed to make maximum use of daylight. The bulbs should be cleaned to remove dust and thus obtain maximum light. Task lights can be used to reduce energy consumption also
- Refrigerator: Keep the back side of the refrigerator in an area where direct sunlight or hot air is not present. Hot food should not be stored in the refrigerator. Set the cooling level of the items in the deep freezer to the lowest level
- Washing machines: reduce the number of washing cycles needed by washing several clothes at one time. Refrain from washing loads of clothing larger or smaller than the approved quantities. Front-loading machines should be used instead of top-loading machines as they have a higher efficiency
- Fans: try to use electric fans instead of air conditioners. Use curtains or grow trees if direct sunlight comes into an air-conditioned room

- Air conditioners: Windows and doors should be closed properly when using an air conditioner. Always maintain the temperature at 26°C, as each reduction of 10°C increases cost by 6%
- Water pumps: use an appropriate water pump when pumping water from a well to a water tank and use rainwater as an alternative to groundwater

3.2.9 Input Material Changes

Input material changes help to minimize wastes and maximize resources at the design stage of a manufacturing process. Specifically, the generation of toxic by-products and environmental impacts can be reduced through input changes (Ganzer *et al.*, 2017). Recycling, reuse, and material recovery processes can also be improved. Different strategies can be used to change input materials. Some of these strategies are:

- Fitting intended use: in this strategy, products and packages are designed to meet optimum use. Durable packaging materials are made to reduce waste generation
- Changing to less hazardous materials: hazardous materials can be replaced by a less hazardous or non-hazardous material. This may reduce the cost of treatment processes needed to meet environmental regulatory limits
- Improved catalysts can be used to enhance the product yield

3.2.10 Process/Equipment Changes

Production process efficiency can be improved to significantly reduce waste generation at the source. This can be achieved by improving current operation and maintenance procedures, modifying existing equipment, and changing the materials used in production (Barnett and Clark, 1996).

3.2.10.1 Improving Operation Procedures

Improved operation procedures help to optimize the use of raw materials in the production process. These methods are usually inexpensive to institute and involve little or no capital expenditure.

3.2.10.2 Equipment Modification

More efficient equipment can be installed to reduce waste generation. This provides many advantages, such as a better production technique. Further, this process reduces the defects in the production process, and can directly reducte economic losses.

3.2.10.3 Material Changes

Process materials can be changed to increase the efficiency of the production process and to reduce waste generation. Also, by reducing the number of rejected or off-specification products, high efficiency systems can reduce the amount of material that must be reworked or discarded. Relatively simple and inexpensive alterations can help ensure that materials are not wasted, leaked, spilt, or lost.

3.2.11 Pollution Control

Environment pollution encompasses all stages of a product or service's life cycle, starting with raw materials and ending with waste treatment. The pollution can be a form of air, water, or soil pollution (Sirait, 2018). The rapid consumption of raw materials and energy, the burning of fossil fuels, deforestation, emissions from industries, land-use pattern changes, etc. can all lead to pollution and result in detrimental environmental issues. Clean technologies and processes help to reduce emissions, conserve resources and energy, and manage waste;treatment technologies provide better options to reduce pollution. There are different practices used for emission reduction in various industries (Sirait, 2018).

3.2.11.1 Air Emission Control

Substances found in the atmosphere that have harmful effects on humans, animals, plants, and the environment are considered air pollutants. Air pollutants are emitted into the atmosphere by anthropogenic activities such as industrial processes, fossil-fuel burning, and volatile compound usage, and by naturally occurring phenomena, such as dust storms and volcanic eruptions (Steiner, 2013). Solid particulate and liquid or aerosol mists, as well as invisible, gaseous-type compounds, are considered the major types of air pollutants. Control technologies should be selected according to the physical state, chemical composition, type, and sources of the pollutant (Steiner, 2013).

Pollution prevention approaches such as using less toxic raw material, improving process efficiency, and using renewable energy sources all help to reduce, eliminate, or prevent pollution. Control technologies such as wet scrubbers, mechanical collectors, electrostatic precipitators, fabric filters (bag filters), condensers, absorbers, combustion systems (thermal oxidizers), adsorbers, and biological degradation can be used in industries (Steiner, 2013).

3.2.11.1.1 Controlling Emissions Related to Transportation

Controlling emissions related to transportation can include emission controls for vehicles as well as the use of cleaner fuels.

Different technological advancements are used to control automobile emissions throughout the world. Many scientific endeavors and innovations are formed to control emission through modifying the vehicles, altering the fuel sources, and increasing fuel-use efficiency. The following techniques can be used to reduce automobile emissions in companies:

- Regular tune-ups are important to maintain vehicles in a proper manner. The manufacturer's maintenance schedule should be used during the maintaining process. The recommended motor oil should be always used, and the move to low-cost, low-quality fuel should be avoided
- The Green Vehicle Guide, which was developed by the EPA, should be used to manage the transportation activities in a company. Specifically, the guide provides a brief on how to use vehicles efficiently and how to reduce pollution by utilizing vehicles powered by alternative fuels such as electricity, hydrogen fuel cells, and cleaner-burning gasoline

- The unnecessary idling of vehicles should be minimized to reduce air pollution, fuel wastage, and excess engine wear. Vehicles engines should not be turned on until they are ready to be driven
- Alternative fuel sources (fuels other than petrol or diesel) should be used to run vehicles, and alternative technologies that do not solely involve petroleum should be used to power the engine. Electric cars, hybrid electric vehicles, and solar-powered vehicles are the best examples of alternative fuel vehicles. There are many advantages to these vehicles, including:
  - Low environmental emissions
  - Best solution for high oil prices and non-renewable fuel usage
  - High efficiency

By increasing the demand for these vehicles, their manufacturers become focused on developing more advanced technological versions of them. For example, produced and sold all around the world, there were:

- 55 million flex fuel automobiles, motorcycles, and light duty trucks by 2015
- 22.7 million natural gas vehicles by 2015
- 24.9 million LPG powered vehicles by December 2013
- 12 million hybrid electric vehicles by 2016
- 1.22 million neat-ethanol-only, light vehicles by 2011
- 2 million highway-legal, plug-in electric, passenger cars and light utility vehicles by 2016

There are different types of vehicles manufactured as alternative fuel vehicles in the world, including: battery electric vehicles (BEVs), solar cars, ammonia-fueled vehicles, bio-alcohol and ethanol–ethanol vehicles, biodiesel vehicles, and biogas vehicles (Shaheen and Lipman, 2007).

- Battery electric vehicles (BEVs)/all-electric vehicles (AEVs): these are known as electric vehicles. The chemical energy of batteries is used as the main energy storage for these vehicles. The California Air Resources Board (CARB) termed BEVs "zero emission vehicles" (ZEV), since the operation process does not produce any tailpipe emissions. Generally, lead-acid batteries and NiMH batteries were used by battery powered cars
- Solar cars: solar panels that are attached or mounted on the cars are used to generate electricity and power the vehicles. Currently, the total power required to fully support the vehicles cannot be supplied, but solar energy can be used to extend the range of electric vehicles
- Ammonia-fueled vehicles: an internal combustion vehicle can use ammonia as an alternative fuel source to fossil fuels. Ammonia engines or motors can, and occasionally are, used on these vehicles
- Bio-alcohol and ethanol vehicles: alcohol, ethanol, or methanol are obtained from petroleum or natural gas, have been used as an alternative fuel source.

Ethanol is considered a renewable resource since it can be extracted from sugar or starch in crops, and other agricultural produce such as grain, sugarcane, sugar beets, or even lactose. Ethanol can be extracted easily where yeast reacts with a sugar solution, such as in overripe fruits. Modern cars designed to operate on gasoline are capable of operating with a blend of ethanol (10–15%) and gasoline (E10–E15). Gasoline-powered vehicles can be modified to run on high ethanol concentrations (85%, or E85). Specifically, the United States and Europe use high-ethanol concentrations in winter seasons and Brazil uses 100% ethanol (E100) in warmer climatic conditions

- Biodiesel vehicles: diesel combustion engines are higher in efficiency (25–30%) than gasoline engines. Diesel that is extracted from biomass is called biodiesel and is commercially available in most oilseed-producing states in the United States. Biodiesel vehicles are unable to keep up with the fuel economy since the energy density of biodiesel is lower than fossil diesel. However, biodiesel produces less emissions than fossil fuel sources because biodiesel contains more oxygen than diesel or vegetable oil fuel
- Natural gas vehicles (NGVs): natural gas mainly consists of methane that it is highly compressed for use as an alternative fuel source to gasoline combustion engine. This highly compressed methane is known as compressed natural gas, or CNG. Methane combustion produces less CO_2 than fossil fuels and this helps to reduce the environmental impacts and carbon footprint of an organization. Modified gasoline cars with CNG and gasoline tanks are called biofuel NGVs, and they can be operated according to their users' preferences. A special option is available to switch between gasoline and CNG while driving these vehicles. In Italy and New Zealand, where natural gases are abundant, NGVs are popular and used widely
- Biogas vehicles: raw biogas is purified and compressed for use as a fuel source for internal combustion engines. After removing the H_2O, H_2S, and particles of the biogas, it is of the same quality as CNG
- Charcoal-powered vehicles: Tang Zhongming pioneered the use of charcoal resources for the Chinese automobile market in the 1930s. The charcoal-fueled car was later used extensively in China, serving the army and conveyancer after the breakout of World War II
- Formic acid vehicles: formic acid is used in fuel cells by converting it to hydrogen, since the storage process of formic acid is much easier than that of hydrogen. In these vehicles, hydrogen is used as the primary energy source for locomotion, and there are two methods to generate power: fuel cell conversion or combustion
- Liquid nitrogen car: liquid nitrogen (LN_2) is a technique for storing energy. Energy is used to liquefy air, and then LN_2 is produced by evaporation and distributed as the fuel source. The LN_2 is exposed to ambient temperatures in the car and as a result nitrogen gas is produced, which can be used to power a piston or turbine engine. The maximum amount of energy that can be extracted from LN_2 is 213 Watt-hours per kg (W·h/kg) or 173 W·h per liter, of which a maximum of 70 W·h/kg can be utilized with an isothermal expansion process

- Liquefied natural gas (LNG): cryogenic liquid is formed by cooling the natural gas to a very low temperature. The LNG supply chain is equivalent to the supply chain of diesel or gasoline. First, natural gas is liquefied in huge amounts; this is similar to the purifying process of gasoline or diesel. Then, using a semi-trailer, the LNG is transported to fuel stations where it is stored in bulk tanks until distributed into vehicles
- Autogas (LPG): LPG can be used to retrofit gasoline cars and these vehicles can then be operated as biofuel vehicles by using both the LPG tank and the gasoline tank together. Approximately 10 million vehicles in the world are operating as autogas vehicles
- Steam cars: A steam engine is used to operate these vehicles. This steam can be produced using different fuel sources such as wood, coal, ethanol, etc. The heat, which converts water into steam, is produced by burning these fuel sources. Steam power can be combined with a standard oil-based engine to create a hybrid
- Wood gas: wood gasifiers should be attached to vehicles to enable the use of wood gas as the fuel source for ordinary internal combustion engines. Several European and Asian countries used these vehicles during World War II because an easy and cost-effective way mode was needed to assess oils in the war
- Hybrids: hybrid vehicles are gaining attention all around the world since they are environmentally friendly and very efficient vehicles. The power required by the vehicle is produced using multiple propulsion. Gasoline–electric hybrid vehicles, which use gasoline (petrol) and electric batteries for energy production, are the most popular in the world, with internal-combustion engines (ICEs) and electric motors used to generate power in these vehicles

The following are two kinds of transportation management strategies:

- Intelligent transportation systems (ITS): The ITS is a novel technology used to control emissions and increase the efficiency of the transportation sector. It aims to provide innovative services linked to different modes of transport and traffic management. With it, users would be better informed, more coordinated, and safer, and there would be a "smarter" application of transport networks. A graphical user interface would be used in ITS to display highway networks, road maps, traffic, and data points, which may refer to all modes of transport. On the 7 July 2010, the European Union defined ITS as systems "in which information and communication technologies are applied in the field of road transport, including infrastructure, vehicles, and users, and in traffic management and mobility management, as well as for interfaces with other modes of transport". ITS can be used to improve the efficiency of the transportation sector with respect to a number of conditions such as road transport, traffic management, mobility, etc. (Ozbay and Kachroo, 1999)
- Mobility management (MM) strategies: MM is a methodology, related to emerging social consciousness around environmental action, with the aim of improving the effect of current traffic policy that influencing individual

awareness. Some psychological factors are used to encourage voluntary behavior alterations in vehicle use. Specifically, transportation demand management (TDM), specific information on public transportation, travel campaigns development, and travel education programs are included in the MM technique. Travel behavior changes should focus on reporting travel behaviors, studying the carbon emissions of travel, calculating the CO_2 emission of travel, providing suggestions to reduce emissions, and advising on efficient driving techniques. Organizations can focus on these awareness programs to improve the efficiency of their transportation sectors while reducing the emission. Specifically, employees can be encouraged to use public transportation (Kim *et al.*, 2017)

Alongside these techniques, an organization can also use low-speed modes, park-and-ride facilities, and integrated regional smart cars, as well as smart growth, parking cash out, telecommuting, and carpooling techniques to reduce the emissions of their transportation sector (Shaheen and Lipman, 2007).

Economic incentives, such as emissions trading, banking, and emissions caps, can also be used. These strategies may be combined with the "command-and-control" type regulations that have traditionally been used by air pollution control agencies.

3.2.11.2 Wastewater Management

Different technologies are used to treat industrial wastewater produced by industries as an undesirable by-product. The treated effluent can be released into the environment or used for industrial processes (Kumar *et al.*, 2017). Table 3.4 shows different technologies used in wastewater treatment.

3.2.11.3 Solid Waste Management

Solid waste management is gaining increasing attention in industries because solid waste generation creates many negative consequences for the environment as well as the economy (Kanchana *et al.*, 2014). Industries produce solid waste in the form of food wastes, housekeeping wastes, packaging wastes, ashes, medical wastes, construction waste, and other hazardous wastes (Kanchana *et al.*, 2014). Different negative impacts, such as the spread of diseases, odor formation, soil pollution, ground water pollution, etc., are caused by solid wastes (Kanchana *et al.*, 2014). Different technologies and methods are used for solid waste management as follows:

- Sanitary Landfill: waste is spread out in thin layers, compressed, and covered with soil or plastic foam. Filled solid waste layers are covered with layers of sand, clay, topsoil, and gravel to prevent the seepage of water
- Incineration: solid wastes are burned at high temperatures until the wastes are turned into ashes. The volume of waste is reduced by up to 20–30% of the original volume
- Recovery and recycling: these processes reduce energy loss, raw material consumption, and landfills
- Composting: biodegradable solid wastes are allowed to decompose in a medium; the result is organic fertilizers for agriculture

TABLE 3.4
Different Technologies Used in Wastewater Treatment

Water Pollutant	Treatment Methods	Industry
Solids (waste product, organic materials, and sand)	Filtration, ultrafiltration, flocculation, primary sedimentation, dissolved air flotation, belt filtration (micro screening), drum screening	Food industry
Salt ions	Brine treatment (reverse osmosis, electrodialysis or weak acid cation exchange, evaporation)	Food and beverage industry
Oils and grease	Dissolved air flotation, media filters, hydrocyclones, oil separators	Electric power plants, food and beverage industry
Biodegradable organics	Activated sludge process, trickling filter	Food and beverage industry, milk industry
Organics (solvents, paints, pharmaceuticals, pesticides)	Advanced oxidation processing, distillation, adsorption, vitrification, incineration, chemical immobilization	Food and beverage industry, organic chemicals manufacturing industry
Acids and alkalis	Neutralized under controlled conditions	Iron and steel industry
Toxic materials (Organic materials, zinc, silver, cadmium, thallium, acids, alkalis, non-metallic elements e.g. arsenic, selenium)	Landfilling, recycling, incineration, molecular encapsulation	Battery manufacturing industry

- Pyrolysis: solid wastes are chemically decomposed by heat (up to 430°C) without the presence of oxygen. The solid wastes change into gasses, solid residue, and small quantities of liquid

3.2.12 Design for the Environment

Design for the Environment (DfE) is an approach to reducing the negative impacts of the life cycle of a product, process, or service. The optimum levels of products or processes/services can be analyzed using different software (Fitzgerald *et al.*, 2007). DfE eliminates the hazardous material usage in raw material extraction, processing. and manufacturing stages to ensure the health of the environment and of employees. DfE includes different techniques aimed at the minimization of waste and hazardous by-products, as well as air pollution, energy expenditure, and other factors (Fitzgerald *et al.*, 2007).

- A design for environmentally friendly manufacturing involves:
  - Non-toxic processes and production materials
  - The minimization of emissions

- Minimum energy utilization
- The Minimization of waste, scrap, and by-products
- A design for environmentally friendly packaging involves:
  - Minimum packaging materials
  - Recyclable packaging materials
  - Reusable pallets, totes, and packaging
  - Biodegradable packaging materials
- A design for better disposal and recyclability involves:
  - Reuse/refurbishment of components and assemblies
  - Avoiding filler material in plastics
  - Minimizing toxicity
  - Selection of materials that enable re-use
  - Identification materials that facilitate re-use
  - Minimum number of materials
  - Designs that enable materials to be easily separated
- A design for disassembly should:
  - Avoid the use of adhesives
  - Limit contaminants
  - Maximize use of recycled or ground material with virgin material
  - Design modular products that can be disassembled for service or re-use
  - Minimize weight of individual parts and modules to facilitate disassembly
  - Provide ready access to fasteners, parts, etc. to support disassembly
  - Minimize fragile parts and leads to enable re-use and re-assembly
  - Use connectors instead of hard-wired connections
  - Use techniques to facilitate disassembly
  - Design to enable the use of common hand tools for disassembly

3.2.13 Life Cycle Assessment

The term life cycle refers to the idea that a fair, holistic assessment involves raw material production, manufacture, distribution, use, and disposal, and includes all intervening transportation steps essential or caused by a product's life.

Life cycle assessment (LCA) is a technique for assessing the total environmental impacts related to all stages of a product's life cycle (from the raw material extraction to disposal stage) (Anex and Lifset, 2014). The full range of environmental effects associated with a product's life cycle can be compared using LCA. is the assessment quantifies all processes and assesses the environmental impacts of them by considering environmental standards such as ISO 14040 and ISO 14044 (Anex and Lifset, 2014).

The main objective of LCA is to compare the full range of environmental effects caused by products and services, and to quantify all impacts by considering all inputs and outputs of material flows. These material flows are assessed to identify environmental impacts. The information from the assessment is used to improve production processes and support policy, and to provide a sound basis for informed decisions.

There are two main types of LCA: attributional LCA and consequential LCA:

- Attributional LCA: this concept estimates the attributes of burdens that are associated with the production or usage of a specific service or process at a point in time
- Consequential LCA: this concept identifies the environmental consequences of a decision or a proposed change in a system that is future-oriented by considering the market and economic implications of a decision

The GHG product life cycle assessments are used to comply with specifications such as PAS 2050 and the GHG Protocol Life Cycle Accounting and Reporting Standard. LCA is used to divide out the environmental load of a process when a number of products or functions share a similar process. Generally, this allocation can be done by using three different techniques: system expansion, substitution, and partition. These techniques are not easy to implement and various methods may lead to different results.

LCA consists of four mandatory elements as follows:

1. Goal and scope definition
2. Inventory analysis
3. Impact assessment
4. Interpretation

3.2.13.1 LCA Goal and Scope Definition

In LCA, a clear goal and scope should be developed for the assessment. For this purpose, a system boundary and functional unit should be selected. A specific goal should be designed by considering the selected functional unit and system boundary. The goal and scope directly impact on results and should be carefully defined for the LCA study.

- The system boundary: this describes the processes or life cycle stages that are included in the LCA. Many assumptions and limitations should be made to partition the environmental load of a process when several products or functions share the same process. The impact categories chosen might include human toxicity, smog, global warming, eutrophication (Anex and Lifset, 2014).
- The functional unit: this defines what is being studied and quantifies the service provided by the product system, providing a reference that the inputs and outputs can be related to. Further, the functional unit is an important basis that enables alternative goods, or services, to be compared and analyzed (Curran, 2013).

3.2.13.2 Life Cycle Inventory (LCI) Analysis

In this step, an inventory of all the environmental inputs and outputs associated with a product or service is created and taken into calculations. Inventory flows include the consumption of water, energy, and raw materials as inputs, and air, land, and water emissions as outputs. A flow model can be developed using a flowchart to illustrate all activities, inputs, and outputs in order to give a clear picture of the technical system boundaries (Curran, 2013).The data from the LCI must be integrated into the functional unit defined in the goal and scope phase. Tables can be used to represent the data and some interpretations can be made at this stage. The completed inventory provides information on all inputs and outputs in the production flow that are related to the environment. The system boundary will decide the amount of information that should be taken into account in the inventory. Generic or brand-specific level information is collected for product LCAs using a questionnaire survey. An industry-level LCA should be carefully monitored to ensure that questionnaires are completed by a representative sample. Producers, sellers, buyers, transporters, etc. should be included in the sample to ensure a balanced assessment procedure. Specifically, best- or worst-case scenarios should not be included in the information gathered on the selected process. Regional differences, social factors, resource usage, and other consumption patterns should be taken into account. The questionnaires should cover all inputs and outputs, typically aiming to account for and calculate 99% of the mass of a product, 99% of the energy usage in its production, and any environmentally sensitive flows, even if they fall within the 1% level of inputs.

Data collection is considered a difficult process within the anthropogenic environment/technosphere. The technosphere is defined as a man-made world and is taken into account by geologists as a secondary resource. Resources that are taken from technosphere are 100% recyclable; resource recovery is considered the main objective of this concept. Man-made resources used in the production process are hard to specify and their end-use is difficult to calculate. Their input is also hard to define, and it is difficult to collect information for LCI. Characteristically, they will not have access to information relating to the inputs and outputs for previous production processes of technospheric material. If it does not possess the relevant information from its own previous studies, the LCA team must then turn to secondary sources. Secondary information should be reflect the regional and national conditions of its own production process. Usually, a standard national database that consists of relevant information on a product should be developed in the country for use in LCA-practitioner tools.

LCI follows different methods to create an inventory, and those are LCI computation using process flow diagrams, matrix expressions of product systems, input–output-based (IO-based) LCI, and the hybrid approach. The hybrid approach can be categorized into three different forms: tiered hybrid analysis, IO-based hybrid analysis, and integrated hybrid analysis.

3.2.13.3 Life Cycle Impact Assessment (LCIA)

LCIA provides the best alternatives and helps to make decisions in the organization. Environmental impacts are classified and evaluated according to their importance and severity. All impacts are described using major impact categories such as global warming, acid rain, or human health.

An impact assessment is formed just after completing the inventory analysis. In this phase, the potential environmental impacts of a product or service are evaluated by considering the results of the life cycle inventory. A typical LCIA phase requires the selection of three mandatory elements: impact categories, category indicators, and characterization models. There are also some optional elements: normalization, grouping, and weighting. Impact categories are selected according to the inventory results. Inventory parameters are sorted and assigned to the specific impact categories. The degree of impact is taken into account at this stage and impacts are ranked according to their severity. All impact categories are characterized into common equivalence units and then those values are summed to produce the overall degree of impacts caused by a product or service. According to ISO 14044:2006, LCIA characterization is considered the final compulsory stage of LCA and marks the conclusion of the study. Optional LCIA elements may be conducted in LCA but this depends on the defined goal and scope of the study. After categorizing the impacts, results can be normalized. The impact categories are usually compared with the total impacts of the region/country. Different environmental impacts are weighted relative to each other in the weighting phase and are then summed to get a single amount for the total environmental impact.

Life cycle impacts can be categorized under the several phases of the lifecycle of a product or a service. For example, development, production, use, and disposal phases can be taken into account. Impacts can then be categorized as first impacts, use impacts, and end-of-life impacts.

- First impacts:
  - Extraction of raw materials
  - Manufacturing (conversion of raw materials into a product)
  - Transportation of the product to a market or site
  - Construction/installation
  - The beginning of the use or occupancy
- Use impacts:
  - Physical impacts of operating the product or facility (such as energy, water, etc.)
  - Maintenance
  - Renovation
  - Repairs (required for continuing to use the product or facility)
- End-of-life impacts
  - Demolition
  - Processing of waste
  - Recyclable materials

3.2.13.4 Interpretation

Interpretation is the final stage of LCA and is a systematic technique used to identify, quantify, check, and evaluate the information from the results of the life cycle inventory and/or the life cycle impact assessment. The inventory analysis and impact assessment results are summarized during the interpretation phase. The outcome of the interpretation phase is a set of conclusions, validations, and recommendations for the study. According to ISO 14040:2006, the interpretation should include:

- Identification of significant problems based on the outcomes of the LCI and LCIA phases of an LCA
- Evaluation of the study with respect to completeness, sensitivity, and consistency checks
- Conclusions, limitations, and recommendations

A key objective in performing life cycle interpretation is to decide on a level of confidence in the final results and to transfer the results in a reasonable, complete, and precise manner. Interpreting the results of an LCA is a challenging step; it starts with understanding the precision of the results, and confirming the results meet the goal of the study. This step is accomplished by identifying the information elements that contribute to each impact category. The sensitivity of the data elements is evaluated and the completeness of the LCA study is assessed. Also, the consistency of the study is evaluated to validate the results. After validating the results, conclusions are made. Limitations are highlighted while providing suggestions and recommendations for future improvements in the product or service life cycle.

More specifically, LCA provides the best means of evaluating the real cradle-to-grave, negative environmental impacts of a product or service with respect to atmosphere, lithosphere, and hydrosphere.

3.2.13.5 Uses of the Life Cycle Assessment Process

- According to a previous study in 2006, LCA is used for different purposes such as:
  - To support business strategy (18%)
  - To aid the research and development process (18%)
  - As an input for process or product design (15%)
  - For educational purposes (13%)
  - To label products or for product declarations (11%)
- LCA is used in the building and construction sector as an assessment tool for measuring environmental impacts. It provides guidelines for the planning and designing of proposed buildings. Specifically, LCI data can be used to collect basic information on the environmental impacts of processes and materials. The building sector can select the most environmentally friendly products for their operations. As an example, the European ENSLIC Building Project guidelines use LCA as their guidance
- LCA is used in house or commissioning studies and is a very popular approach. The process is supported by the government through providing the facilities for developing national databases. These databases can be used to complete those LCA stages where secondary information is required
- Environmental Product Declarations or the ISO Type III labeling technique is supported by LCA. Environmental product declarations are defined as "quantified environmental data for a product with pre-set categories of

parameters based on the ISO 14040 series of standards, but not excluding additional environmental information". This is a third-party certification process in which the relative environmental merits of competing products can be evaluated through LCA-based labels. It plays a major role in industry and provides a progressively significant source for assessment

- LCA also plays a major role in environmental impact assessment
- LCA is used to assess integrated waste management
- Pollution studies are supported by LCA. For example, a recent study used LCA in a laboratory evaluation of a plant for O_2-enriched air production together with its economic estimation in a holistic eco-design perspective

  LCA has been used to assess the environmental influences of roadway and pavement maintenance, repair, and rehabilitation activities.

3.2.13.6 Data Analysis in the Life Cycle Assessment Process

The whole LCA process depends on the data/information that is used in the studies. Data validation is important to ensure the reliability of the study. Therefore, LCA is built on having accurate, relevant, and current information. LCA is used to compare different products with each other. This process can only be applied when both life cycle assessments use equivalent data for their assessment process. If one assessment used precise data and the other used low-quality data this comparison could not take place.

There are two basic types of data are used in LCA:

- Unit process data
- Environmental input–output data (EIO)

Unit process data originates in direct surveys by companies or plants. This data is collected by considering the system boundaries and unit process levels defined by the studies. EIO data originates from national economic input–output data.

Data validity is of ongoing concern in life cycle analyses. Due to globalization and the rapid development of research, companies are continuously introducing new materials and manufacturing methods to the market. New commodities arriving on the market make the LCA process more difficult since it is hard to gather up-to-date information for the assessment. Data should be validated and recent in order to ensure an accurate result in the assessment; as a result, data collection consumes more time. Ongoing data collection can be implemented, and information can be updated in a timely manner to ensure there is precise, recent data available to make the LCA process easier to validate.

The LCA process measures the environmental impacts of all stages of a product or service's life cycle, including raw material extraction, manufacturing and processing, product use, and product disposal. According to the results of the LCA, an organization can identify the most harmful stages of a product or service's life cycle and take precautions or controlling measures to reduce these environmental impacts. Processes, materials, and products can be modified to reduce these impacts, and more environmentally friendly products or services can be produced. For example, strategies like reducing the demand for energy and other resources, using renewable materials, opting for alternative transportation to reduce emissions, choosing less toxic materials, etc. can be used for the modification of manufacturing processes and to make them more efficient. The decrease throughout the use phase should be more than enough to maintain the stability of additional raw material or manufacturing costs.

Large databases are typically called data sources. Two types of products are not appropriate for comparison if different data sources have been used to source the data. Different types of data sources are available and some are listed below:

- Agribalyse
- Agri-footprint
- BioEnergieDat
- Comprehensive Environmental Data Archive (CEDA)
- Ecoinvent
- Soca
- EuGeos' 15804-IA
- ESU World Food
- GaBi
- LC-Inventories.ch
- NEEDS
- Ökobaudat
- ProBas
- PSILCA
- Social Hotspots
- USDA

Impact can be calculated manually, but generally different software is used to calculate the impacts of a product in the LCA process. This software can range from a simple spreadsheet, where the user enters data manually, to a fully automated program, where the user is not aware of the source data.

3.2.13.7 Types of Life Cycle Assessment Process

There are different types of LCA processes that can be implemented. A defined system boundary will decide on the number of stages in a product or service life cycle that should be accounted. According to the system boundary different LCA types are classified.

3.2.13.7.1 Cradle-to-Grave Analysis

Cradle-to-grave refers to the complete life cycle assessment process of a product or service from raw material extraction ("cradle") to use and disposal phase ("grave").

For example, paper is produced from a tree, and this can be recycled to produce low-energy fiberized paper insulation, which can then be sold to homes as energy-saving devices that can last for four years. This saves 2,000 times the fossil-fuel energy used in the paper production process. The cellulose fibers are replaced after 40 years and the old fibers are disposed of using incineration. All inputs and outputs (materials and energy) are measured for all the stages of the life cycle of product.

3.2.13.7.2 Cradle-to-Gate Analysis

Cradle-to-gate is the assessment process of a partial product or service life cycle from resource extraction ("cradle") to end of the production, just before the product is transported to the consumer ("gate"). The final stages of the product life cycle (use and disposal stages) are excluded from the assessment process. Generally, environmental product declarations use a cradle-to-gate assessment process, called business-to-business EDPs. This is a popular type of LCA process since data collection can be conducted easily. Data can be collected by considering all resources that are purchased and used by the facility. Specifically, the LCI phase uses data from raw material extraction to end of production, and it can include all transport to plant and manufacture processes more easily. Organizations can produce their own cradle-to-gate values for their products and services.

3.2.13.7.3 Cradle-to-Cradle or Closed-Loop Production

Cradle-to-cradle is a specific type of LCA process that assess product or service life cycles from raw material extraction ("cradle") to specified disposal process in the form of a recycling process ("cradle"). Recycled materials are used as an input material for a product or service and this is considered a cradle stage. Recycling raw materials helps to reduce many environmental impacts because it minimizes waste generation and virgin material extraction. It also helps to employ sustainable production, operation, and disposal practices, as well as targets to integrate social responsibility into the development of a product or service. The recycling process can creates new products or services identical to the first product or service. For example, glass bottles are created from recycled glass bottles, paper is created from used paper, and asphalt pavements are created from discarded asphalt pavements. Also, the recycling process can create products different to the first product. For example,, glass wool separation from used glass bottles, and the creation of different plastic items such as baskets, bottles, and cups from various used plastics. The closed-loop LCA process is difficult to implement as there are the barriers of data collection and inventory preparation. But, product and service allocation to this process is easier than the open loop LCA process since all materials and services are used in the process are taken into account. The cradle-to-cradle LCA process takes place at the point of product allocation and the system boundary is very difficult to define.

3.2.13.7.4 Gate-to-Gate Analysis

Gate-to-gate analysis, which is also a partial LCA process, measures the impacts created within an organization. Typically, the stages of raw material extraction, product usage, and product disposal are excluded from the assessment. This assessment looks at only one value-added process in the entire production chain. A gate-to-gate assessment process may also later be connected with the appropriate production chain of the organization to form a complete cradle-to-gate evaluation.

3.2.13.7.5 Well-to-Wheel Analysis

Well-to-wheel is a specific LCA process that is used to measure the impacts of transport fuels and vehicles. The analysis is frequently broken down into different stages as follows:

- Well-to-station
- Well-to-tank
- Station-to-wheel
- Tank-to-wheel
- Plug-to-wheel

The feedstock or fuel production and fuel delivery or energy transmission processes are included in the first stage, called the "upstream" stage. The components of the "downstream" stage include the processes in the vehicle. The well-to-wheel analysis is frequently used to measure total energy consumption, energy conversion efficiency, emissions generated by marine vessels, aircraft, and motor vehicles, together with their carbon footprint, and the fuels consumption of each of these transport modes. The well-to-wheel examination is beneficial for highlighting these various emissions and the efficiencies of energy technologies and fuels at both the upstream and downstream stages. The examination provides a more complete picture of actual emissions.

The Greenhouse Gases, Regulated Emissions, and Energy Use in Transportation (GREET) model, which was developed by the Argonne National Laboratory, is used to calculate the impacts of new fuels and vehicle technologies. The well-to-wheel LCA process has a substantial input in this model. The model estimates the impacts of fuel use by using a well-to-wheel assessment while a cradle-to-grave method is used to identify the impacts of the vehicle itself. Energy use, greenhouse gas emissions, and six additional pollutants are reported and evaluated by the model. These pollutants are: carbon monoxide (CO), volatile organic compounds (VOCs), nitrogen oxide (NOx), sulfur oxides (SOx), particulate matter with a size smaller than 2.5 micrometers ($PM_{2.5}$) and 10 micrometers (PM_{10}). Well-to-wheel analysis and LCA differ in that LCA measures all impacts by considering more emission sources than while well-to-wheel analysis, which counts only those emissions caused by fuels. Thus, quantitative values of GHGs can differ between the two analyses.

3.2.13.7.6 Economic Input–Output Life Cycle Assessment

An economic input–output LCA (EIO-LCA) measures the values of combined sector-level information on the amount of environmental impacts that can be linked to each sector of the economy and the amount of purchasing from other sectors that is implied to take place in each economic sector. Long chains of economic processes can be analyzed using EIO-LCA and this leaves the LCA process to define scopes and system boundaries. EIO-LCA is not suitable for measuring the environmental impacts of a product. Furthermore, economic quantities cannot be translated into environmental impacts, and the data collected is not validated.

3.2.13.7.7 Ecologically Based LCA (Eco-LCA)

Eco-LCA was developed to provide direction on the wise management of anthropogenic activities by considering the direct and indirect impacts of these activities on natural resources and surrounding ecosystems. Thus, Eco-LCA provides a broad measure of environmental impacts. Conventional LCA and Eco-LCA use similar approaches and strategies to evaluate these impacts. Eco-LCA was designed by the Ohio State University Center for Resilience and is used to quantify the impacts of regulating and supporting services that are applied in the life cycle of economic goods and products.

The services are categorized into four main groups by Eco-LCA as follows:

- Supporting services
- Regulating services
- Provisioning services
- Cultural services

3.2.13.7.8 Exergy-Based LCA

The maximum beneficial work conceivable during a process that carries the system into equilibrium with a heat reservoir is called the exergy of the system. DeWulf and Sciubba have stated the relationship between exergy analysis and resource accounting. The process directive is to exergo–economic accounting and to methods specifically devoted to LCA such as exergetic material input per unit of service (EMIPS). The concept of material input per unit of service (MIPS) is calculated in terms of the second law of thermodynamics, permitting the measurement of both resource input and service output in terms of exergy. EMIPS is gives details on transport technology: the process considers the service together with the total mass to be transported, and the total distance. The mass per single transport and the delivery time also can be taken into account in this analysis process.

3.2.13.7.9 Life Cycle Energy Analysis

Life cycle energy analysis (LCEA) is a technique used to account for all energy inputs into a product by considering direct energy inputs during manufacture together with energy inputs needed to produce components, materials, and services for the manufacturing process. The technique is applied to analyze the total life cycle energy input of a product or a service.

There are other energy-related approaches formed under the LCEA process, such as energy production, energy cannibalism, and energy recovery:

1. Energy Production

   Energy production processes such as photovoltaic electricity, nuclear energy, and high-quality petroleum products waste much energy when producing energy commodities, as seen in Equation (3.4).

$$\begin{aligned}\text{net energy content} &= \text{energy content of the product} \\ &\quad - \text{energy input used during extraction and conversion}\end{aligned} \tag{3.4}$$

Net energy content is calculated using both direct and indirect energy extraction and conversion. An LCEA study claimed that manufacturing solar cells involves more energy than is recovered by the cell.

2. Energy Cannibalism

   The energy demand of existing power plants can increase due to the rapid growth of energy-intensive industries, creating a "cannibalism" effect. This concept is used in the UK to determine the life cycle energy impacts of many types of renewable technologies.

3. Energy Recovery

   Organizations can reproduce energy through the incineration process, if materials are disposed of using the incineration process. The process can harness low-impact energy sources with less emissions. Also, by producing energy, this reduces the high emission of GHGs associated with the incineration process.

3.2.14 Green Purchasing

Green purchasing is defined as the procurement of products and services that have a lesser or reduced effect on human health and the environment when compared with competing products or services that serve the same purpose (Murray, 2000). Green purchasing considers the whole life cycle of a product, from raw material extraction to the disposal stage (that is, raw materials acquisition, production, manufacturing, packaging, distribution, reuse, operation, and maintenance or disposal of the product or service). Green products can save more money than the traditional products because of cleaner and safer production processes. Recycle, reuse, or recovery techniques during the manufacturing process reduce the environmental impacts and cost of production (Green *et al.*, 1998). Green purchasing promotes these green product manufacturing processes and helps to maintain sustainability in the manufacturing industry. As an example, buying 100%-recycled-content paper reduces energy use by 44%, decreases greenhouse gas emissions by 37%, cuts solid waste emissions in half, decreases water use by 50%, and practically eliminates wood usage (Murray, 2000).

3.2.14.1 Benefits of Green Purchasing

- Improved brand image among consumers
- Increased customer satisfaction
- Reduced risk of green regulations, environmental impacts, hazardous compound formation, accidents, and waste generation
- Reduction in costs such as hazardous material management costs, operational costs, repair and replacement costs, disposal costs, health and safety costs, and environmental costs all as a result of consuming less energy and less resources, and generating less waste
- Increased shareholder value as a result of the improved brand value, accelerating the market demand of the product

3.2.14.2 Steps of Green Procurement

- Decide to be green: create a green policy in the organization. All procurements of products and services must meet minimum green requirements, and written authorizations must be required to choose a product that is not among the greenest possible choices
- Identify and categorize: identify the areas where waste or pollution can occur or where it is necessary to modify a production process in order to green it
- Develop green specifications and standards and identify the minimum requirements, which are:
  - Highly recyclable: design products for easy disassembly and recycling, and make them returnable to the manufacturer under an end-of-life program that will see that the component is refurbished or recycled
  - IT: purchase energy efficient equipment with a long life cycle and that is free of hazardous materials
  - Repairable: use repairable equipment to prevent the necessary replacement of the whole system
  - Use existing environmental standards such as Energy Star, or standard environmental ratings such as LEED, EMAS, or the ICLEI
- Establish green selection criteria
- Identify products and services that are green, as well as green suppliers and green products for purchasing
- Use a life cycle costing approach: calculate the total environmental impact of a product using life cycle assessments. All environmental impacts, from raw material extraction to the final disposal stage, are must be taken into account
- Include green performance clauses in every contract and terminate the supply of non-green materials to the operation process
- Communicate with and inform all stakeholders about green purchasing practices
- Use green technology, such as e-procurement, e-sourcing, and other e-systems

REFERENCES

Agwu, M.O., and Samuel, O.A. "Good housekeeping - A panacea for slips, trips and falls accident in the NLNG Project, Bonny." *International Journal of Business Administration* 5(4), 2014: 12–20. doi. 10.5430.

Ahuja, I.P.S., and Khamba, J.S. "Total productive maintenance: Literature review and directions." *International Journal of Quality and Reliability Management* 25(7), 2008: 709–756. doi.10.1108.

Al-khatib, Bilal Adel. "The effect of using brainstorming strategy in developing creative problem solving skills among female students in princess alia University College Department of Psychology and special education." *American International Journal of Contemporary Research* 2(10), 2012: 29–38.

Anex, Robert, and Lifset, Reid. "Life cycle assessment." *Journal of Industrial Ecology* 18(3), 2014: 321–323. doi. 10.1111.

Barnett, Brent D., and Clark, Kim B. "Technological newness: An empirical study in the process industries." *Journal of Engineering and Technology Management* 13(3–4), 1996: 263–282. doi. 10.1016.

Bright, D.S., and Newbury, D.E. "Concentration histogram imaging: A scatter diagram technique for viewing two or three related images." *Analytical Chemistry* 63(4), 1991: 243A–250A. doi. 10.1021.

Curran, Mary Ann. "Life cycle assessment: A review of the methodology and its application to sustainability." *Current Opinion in Chemical Engineering* 2(3), 2013: 273–277. doi. 10.1016.

El-Namrouty, Khalil A. "Seven wastes elimination targeted by lean manufacturing case study 'Gaza Strip manufacturing firms'." *International Journal of Economics, Finance and Management Sciences* 1(2), 2013: 68. doi. 10.11648.

Ferreira de Castro, Vinicius, and Frazzon, Enzo Morosini. "Benchmarking of best practices: An overview of the academic literature." *Benchmarking: Anais an International Journal* 24(3), 2017: 750–774. doi. 10.1108.

Fitzgerald, Daniel P., Herrmann, Jeffrey W., Sandborn, Peter A., Schmidt, Linda C., and Gogoll, Thornton H. "Design for environment (dfe): Strategies, practices, guidelines, methods, and tools." *Environmentally Conscious Mechanical Design* 3, 2007: 1–24. doi. 10.1002.

Fonseca, Luis, Lima, Vanda, and Silva, Manuela. "Utilization of quality tools: Does sector and size matter?." ,*International Journal for Quality Research* 9(4), 2015: 605–620.

Freitas, Anelisse Vasco Mascarenhas de, and Quixabeiro, Elinaldo Leite, Luz, Geórgia Rosangela Soares, Franco, Viviane Moreira, and Santos, Viviane Fontes dos . "Standard operating procedure: Implementation, critical analysis, and validation in the audiology Department at CESTEH/Fiocruz." *CoDAS* 28(6), 2016: 739–744. doi. 10.1590.

Ganzer, Paula Patricia, Chais, Cassiane, and Olea, Pelayo Munhoz. "Product, process, marketing and organizational innovation in industries of the flat knitting sector." *RAI – Revista de Administração e Inovação* 14(4), 2017: 321–322. doi.10.1016.

Gomiero, T., and Giampietro, M. "Graphic tools for data representation in integrated analysis of farming systems." *International Journal of Global Environmental Issues* 5(3/4), 2005: 264–301. doi. 10.1504.

Green, K., Morton, B., and New, S. "Green purchasing and supply policies: Do they improve companies\ environmental performance?." *Supply Chain Management: an International Journal* 3(2), 1998: 89–95.

Haefeli, Mathias, Elfering, Achim, McIntosh, Emma, Gray, Alastair, Sukthankar, Atul, and Boos, Norbert. "A cost-benefit analysis using contingent valuation techniques: A feasibility study in spinal surgery." *Value in Health: The Journal of the International Society for Pharmacoeconomics and Outcomes Research* 11(4), 2008: 575–588. doi. 10.1111.

Ibrahim, Rosziati, and Yen, Siow Yen. "Formalization of the data flow diagram rules for consistency check." *International Journal of Software Engineering and its Applications* 1(4), 2010: 95–111. doi. 10.5121.

Johannson, Lynn *A Pocket Guide to Green Productivity, Asian Productivity Organization*, 2005: 1–278. ISBN: 92-833-70457.

Kanchana, S., Kathiravan, R.S., Priyanka, J.., and Neetha Delphin Mary, K. "Industrial solid waste management practices in medium and small scale industries located in TamilNadu." *International Journal of Emerging Technology and Advanced Engineering* 4(6), 2014: 1–6.

Kim, J., Choi, K., Kim, S., and Fujii, S. "How to promote sustainable public bike system from a psychological perspective?" *International Journal of Sustainable Transportation* 11(4), 2017: 272–281.

Liliana, Luca. "A new model of Ishikawa diagram for quality assessment." *IOP Conference Series: Materials Science and Engineering* 161(1), 2016: 1–7. doi.10.1088.

Lymperopoulos, Nikolaos Spyros, Jeevan, Ranjeet, Godwin, Lindsey, Wilkinson, Donnas, James, Malcolm Ian, and Shokrollahi, Kayvan. "The Introduction of standard operating procedures to improve burn care in the United Kingdom." *Journal of Burn Care and Research: Official Publication of the American Burn Association* 36(5), 2015: 565–573.

Ma, F., O'Hare, G.M.P., Zhang, T., and O'Grady, M.J. "Model property based material balance and energy conservation analysis for process industry energy transfer systems." *Energies* 8 (10), 2015: 12283–12303. doi. 10.3390.

Malik, Nur Khaliesah Abdul, Abdullah, Sabrina Ho, and Manaf, Latifah Abd. "Community participation on solid waste segregation through recycling programmes in Putrajaya." *Procedia Environmental Sciences* 30, 2015: 10–14. 10.1016. doi.

Michalska, J., and Szewieczek, D. "The 5S methodology as a tool for improving the organisation." *Journal of Achievements in Materials and Manufacturing Engineering* 24(2), 2007: 211–214.

Murray, J. Gordon. "Effects of a green purchasing strategy: The case of Belfast city council." *Supply Chain Management: an International Journal* 5(1), 2000: 37–44. doi. 10.1108.

Naik, Sanjeev B., and Kallurkar, Shrikant. "A literature review on efficient plant layout design." *International Journal of Industrial Engineering Research and Development* 7(72), 2016: 43–51.

Noll, K.E., Haas, C.N., and Patterson, J.W. "Recovery, recycle and reuse: Of hazardous waste." *Journal of the Air Pollution Control Association* 36(10), 1986: 1163–1168. doi. 10.1080.

O'Rielly, K., and Jeswiet, J. "Strategies to improve industrial energy efficiency." *Procedia CIRP* 15, 2014: 325–330. doi. 10.1016.

Olabanji, O.M., and Mpofu, K. "Comparison of weighted decision matrix, and analytical hierarchy process for CAD design of reconfigurable assembly fixture." *Procedia CIRP* 23(C), 2014: 264–269. doi. 10.1016.

Ozbay, K., and Kachroo, P. *Incident Management in Intelligent Transportation Systems*, 1–248. Norwood: Artech House Publishers, 1999. ISBN: 0890067740.

Rizi, Cobra Emami, Najafipour, Mojtaba, haghani, Faribaani, and Dehghan, Shahla. "The effect of the using the brainstorming method on the academic achievement of students in grade five in Tehran elementary schools." *Procedia - Social and Behavioral Sciences* 83, 2013: 230–233. doi. 10.1016.

Shaheen, S.A., and Lipman, T.E. "Reducing greenhouse emissions and fuel consumption: Sustainable approaches for surface transportation." *IATSS Research* 31(1), 2007: 6–20.

Sharan Kumar, K. Mohammed Suhail, S., and Elango, R. "Water pollution control in paper and pulp industry." *Mperial Journal of Interdisciplinary Research* 3(3), 2017: 466–469.

Sirait, M. "Cleaner production options for reducing industrial waste: The case of batik industry in Malang, East Java-Indonesia." *IOP Conference Series: Earth and Environmental Science* 106(1), 2018: 1–5. doi. 10.1088.

Steiner, B.A. "Air-pollution control in the iron and steel industry." *International Metals Reviews* 21, 2013: 171–192.

Tangkawarow, I.R.H.T., and Waworuntu, J. "A comparative of business process modelling techniques." *IOP Conference Series: Materials Science and Engineering* 128, 2016: 2–17. doi. 10.1088.

Tennant, Ruth, Mohammed, Mohammed A., Coleman, Jamie J., and Martin, Una. "Monitoring patients using control charts: A systematic review." *International Journal for Quality in Health Care* 19(4), 2007: 187–194. doi. 10.1093.

4 End-of-Pipe Treatment Techniques

Guttila Yugantha Jayasinghe,
Shehani Sharadha Maheepala, and
Prabuddhi Chathurika Wijekoon

4.1 WASTE MANAGEMENT

Productivity improvement that is driven by population growth leads to higher rates of waste accumulation on Earth. The potential stress on the Earth's environment is eliminated through waste management strategies.

Waste management is a group of techniques and strategies that are implemented with the primary aim of reducing the environmental impacts of waste generation and disposal. The concept has been integrated into green productivity (GP) with the aim of handling problems associated with pollution and resource conservation (APO, 2004). Waste management plays a vital role in resource conservation by managing the resources that are spent on waste handling and remediation. By integrating waste management techniques into its methodology, GP focuses on the following aspects:

1. Avoid Waste

   In GP, avoiding waste generation can occur simultaneously with productivity improvement. Waste management techniques act as supportive tools in this context, and designs for environment techniques are also supportive in avoiding waste in an organization.
2. Reduce Waste at Source

   Source reduction is a key strategy in waste management that can be applied in various stages of a production process that is incorporating GP. Changing input materials, processes and equipment, and improving operating procedure techniques, may further encourage the source reduction strategy of waste management.
3. Separate Waste at Source

   Waste separation and sorting occurs in institutions where GP is implemented as a primary step for waste management. These activities ease the continuous processes of managing waste and determining its fate. Thus, these activities are helpful in modifying and constructing infrastructure and waste management facilities in an organization during the GP implementation process. Housekeeping activities (such as the 5S organization method) and the Seven Wastes GP technique can be integrated into waste separation.

4. Reuse Waste

   Reusing generated waste can be considered a cost reduction method as well a resource conservation technique in GP. Reusing is one of the key elements of 3R concept, which is categorized as a main GP technique. It can be conceptually developed further as "your waste – another's resource" when reusable waste is granted for use to other organizations or departments.

5. Recycle Waste

   Waste recycling is also an essential focus in GP implementation, and is also included in the 3R concept. By recycling waste, process input material and other valuable products can be generated, which enables resource conservation. The concept of "your waste – another's resource" is further demonstrated by waste recycling.

4.2 END-OF-PIPE TREATMENT

End-of-pipe treatment is one of the most common waste management methods. It is applied at the end of the production process and deals with by-products, with the aim of removing or transforming wastes emitted from the production process. Rather than avoiding or preventing waste generation, end-of-pipe treatment focuses on converting the exiting wastes into harmless substances, allowing for safe disposal that complies with environmental legislations (McCray, 2011).

Manufacturing processes do not have 100% conversion efficiency and this may result in the formation of some wastes in the form of solids, liquids, and gases, as shown in Figure 4.1. Thus, alongside waste avoiding techniques, manufacturers should also consider waste treatment methods at the points where these wastes are generated and released. Thereby, air emission control techniques should be focused on stack emissions, fumes, and odors at the workplace; liquid/effluent treatment plants should be designed for industrial effluents, and domestic and sanitary wastewater, while solid waste management facilities should be catered for industrial solid wastes and sludge from effluent treatment plants.

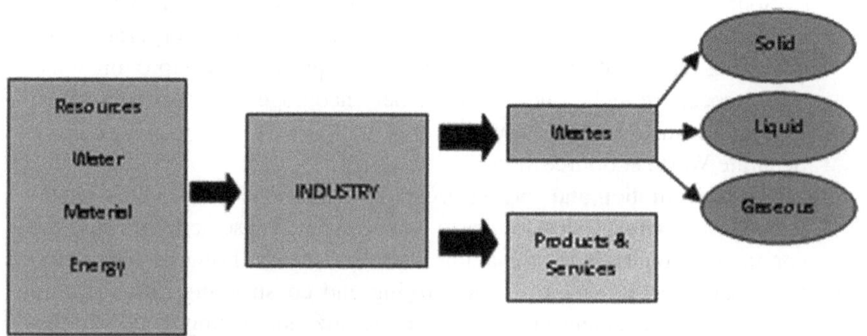

FIGURE 4.1 Waste generation within an organization.

End-of-pipe treatment strategies depend on various aspects related to the wastes produced. According to those aspects, treatment facilities, infrastructure, and disposal methods will vary within a particular organization. Commencing from a single treatment technique, strategies can be widened into a combination of several techniques depending upon those waste aspects.

The aspects to be considered are as follows:

- Type of waste
  The three main types of wastes are:
  - Gaseous
  - Liquid
  - Solid

  Wastes can be further classified into categories by considering the following factors:
  - Hazard potential (hazardous or non-hazardous)
  - Origin of wastes (industrial, domestic, construction, etc.)
  - Content of wastes (organic, metal, e-wastes, etc.)
- Type of treatment
  - Physical
  - Chemical
  - Biological
- Properties of waste stream
  - Chemical
  - Physical
  - Biological

  Dissolved solids, pH, suspended solids, BOD, and COD are indicators of the above properties.

4.2.1 Gaseous Waste Treatment

Production processes contribute to the emission of various gaseous pollutants. Most of combustion reactions result in carbon monoxide, carbon dioxide, unburned carbon particles, and particulate matter ($PM_{2.5}$ and PM_{10}). Impurities in fuel generate gaseous compounds such as sulfur oxides, nitrogen oxide, etc. (Kampa and Castanas, 2008).

Gaseous waste treatment techniques are as follows.

4.2.1.1 Wet Scrubbers

A wet scrubber is a device used to control air pollution by removing particulate matter (PM) and acidic gases that are present in a carrier gas stream. The device uses a liquid, typically water or a water-based solution, to capture the pollutants. The basic processes of impaction, diffusion, interception, and/or absorption are applied in the process of removing the pollutant. The gas stream that is then emitted from the scrubber no longer contains high residual amount of the pollutants and poses no threat of pollution. Pollutant-containing liquid is collected for disposal, undertaking further treatment if required as shown in Figure 4.2. There are numerous types of wet scrubbers that remove both acidic gas and PM (Schifftner and Hesketh, 1996).

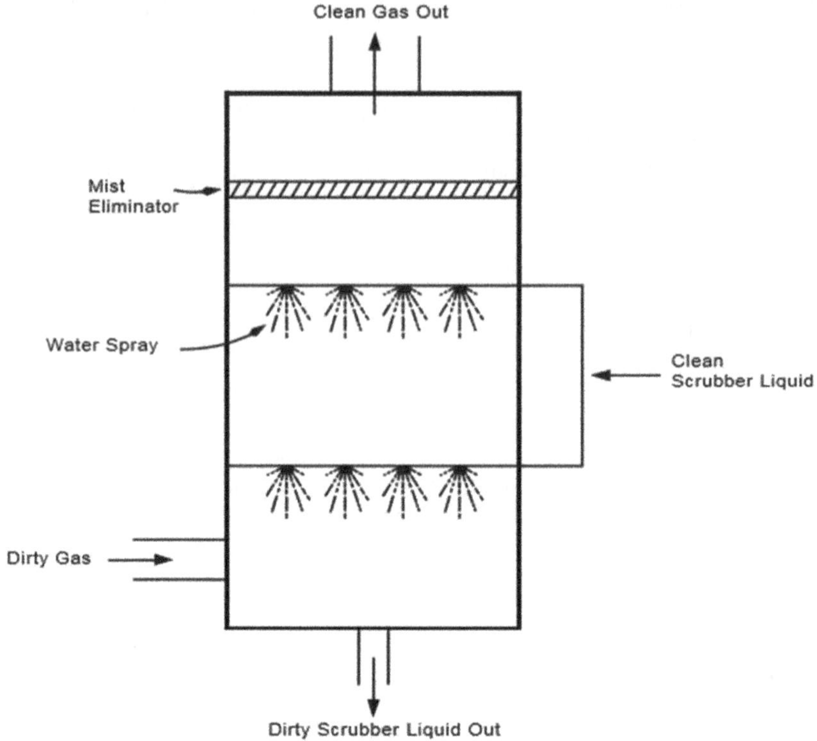

FIGURE 4.2 Basic structure of a wet scrubber.

4.2.1.1.1 Wet Scrubbers for Controlling Particulate Matter

Impaction is the primary capture mechanism among the several mechanisms of contacting liquid droplets with particulates. Waste gas flows along the streamline and around the droplet, where it comes contact with the water droplet. Particles with diameters greater than 10 μm are generally settled due to their mass, while particles with sufficient inertial force maintain their forward route. Turbulent flow enhances capture by impaction (EPA, 1997).

Particles with diameters between 0.1–1.0 μm are subjected to interception while much smaller sized particles are subjected to Brownian motion. Interception is the capturing mechanism due to the surface tension of the water droplet when the particles are passing sufficiently close to it (Schnelle and Brown, 2002). Interception efficiency is enhanced by increasing the density of droplets in a spray.

The collection efficiencies of wet scrubbers vary according to factors such as:

- The particle size distribution of the waste gas stream: collection efficiency decreases where the size of PM decreases
- Scrubber type: collection efficiencies range from greater than 99% for venturi scrubbers to 40–60% (or lower) for simple spray towers. Wet scrubber designs have been improved to obtain increased collection efficiencies

According to differences in mechanism and arrangement, there are several types of scrubbers, including:

1. Spray Tower

   The simplest type of scrubber is the spray tower, in which air containing particulate matter passes into a chamber where nozzles spray it with liquid. Inside the chamber, the air comes in contact with the liquid spray and can occupy either vertical or horizontal waste gas flow paths. The liquid spray can be maintained either in counter direction to the gas flow, in the same direction as the gas flow, or in a position perpendicular to it. The pollutant-containing gas flow is allowed to enter at the bottom of the tower and flow upwards, while nozzles spray water downward. The nozzles can be mounted on the walls of the tower; in some cases, they are placed on an array at the center of the tower. Specific mechanisms of impaction, diffusion, and interception stimulate the arresting of suspended particles by water droplets.

   Larger droplets settle as a result of gravity and collect at the bottom of the tower. The mist eliminator, is located in the chamber, collects the particles that remain entrained in the gas stream. Generally, spray towers exhibit high collection efficiencies for coarse particles. Typical removal efficiencies vary according to the particle size distribution of the waste gas stream, as shown in Table 4.1.

   Spray towers are appropriate for the removal of particulate matter from the air waste in grinding mills, fertilizer plants, asphalt plants, and pigment operations. In comparison to the other wet scrubbers, spray towers have lower capital costs as well as lower operational costs.

2. Cyclonic Spray Tower

   The waste gas stream produces a cyclonic motion inside the chamber which differs from spray tower designs. The cyclonic motion is maintained either by positioning the gas inlet tangential to the wall of the chamber or by placing turning flips within the scrubbing chamber.

   The gas inlet is narrowed as it increases the gas velocity while entering the tower. The scrubbing liquid can be sprayed either from a tangential inlet, where the nozzles are located in a central pipe, or from turning vanes at the top of the tower. Due to the rotational movement, the suspended particles in

TABLE 4.1
The Particle Size Distribution and Removal Efficiencies of Spray Towers

Particle size (μm)	Removal efficiency
Larger than 5	90%
3–5	60–80%
Below 3	Less than 50%

the gas stream undergo a centrifugal force, causing them to move towards the tower walls. Subsequently, they impact with the tower walls, falling to the bottom of the chamber where they are collected. The mist eliminator is located in the chamber, which prevents the remaining entrained droplets in the waste gas stream from escaping.

The droplets in cyclonic spray towers have higher velocities relative to the waste gas. Thus, greater collection efficiencies can be obtained by this arrangement in comparison to simple spray towers. Collection efficiencies of cyclonic spray towers vary as demonstrated in Table 4.2.

Typical applications for cyclonic spray towers are dust control in fertilizer plants, grinding operations, and foundries. Cyclonic spray towers have a more complex design which may result in higher capital, operation, and maintenance costs.

3. Dynamic Scrubber

These scrubbers are similar to spray towers, and are occasionally referred to as mechanically aided scrubbers or disintegrator scrubbers. A power-driven rotor is incorporated into the scrubber to convert the scrubbing liquid into finely dispersed droplets; it can be located inside or outside the tower and is to it connected through a duct. A mist eliminator is located in the dynamic scrubber. A special mechanism of waste gas humidification eliminates evaporation and particle deposition in the rotor area.

Even though dynamic scrubbers efficiently remove fine particulate matter, an additional maintenance cost can be involved due to the rotor. Large PM and humid gas streams can damage the rotors through abrasion corrosion. A cyclone or a pretreatment device may often remove large PM from the waste stream. Power consumption is high for this scrubber while collection efficiencies are similar to cyclonic spray towers. Since a rotor functions in system capital, operation, and maintenance costs are moderately higher than simple spray towers.

4. Tray Towers

A tray tower is a vertical tower that consists of several perforated trays, mounted horizontally inside the tower. The waste gas stream enters at the bottom of the tower while the scrubbing liquid flows from the top. The waste gas stream moves upward through openings in the trays and the scrubbing liquid travels across each tray. The tray arrangement provides more gas–liquid contact, with the gas mixing well with the liquid. The downward

TABLE 4.2
The Particle Size Distribution and Collection Efficiencies of Cyclonic Spray Towers

Particle size	Collection efficiency
Greater than 5 μm	95%
Submicron particles	60–75%

movement of liquid through the perforations in the tray is prevented due to the gas velocity. Tray towers are designed to provide access to each tray for cleaning and maintenance. Although large PM has the potential to clog the perforations, incorporating impingement baffles will avoid gas entering the opening.

Tray towers are not effective in removing submicron particles, although they have collection efficiencies of 97% for particles larger than 5 μm. Tray towers are effective in removing both soluble pollutant gases and particulate matter. The scrubbers are applicable for use in lime kilns, bark boilers bagasse, and secondary metals industries. The capital, operation and maintenance costs of tray and impingement towers are moderately higher than simple spray towers.

5. Venturi Scrubbers

   A "converging–diverging" flow channel is the special feature of venturi scrubbers. The cross-sectional area of the channel first increases and then decreases along the length of the channel creating a narrow area called a "throat". In that area, waste gas velocity and turbulence increases due to the decreasing surface area. The scrubbing liquid is injected directly into the throat section. The high velocity and turbulence atomize the scrubbing liquid and improves the contact of gas and liquid. Then the gas–liquid mixture slows down as it moves through the diverging section where additional particle–droplet collisions and the agglomeration of the droplets can occur. An entrainment section allows for the separation of liquid droplets from the gas stream. This section consists of a cyclonic separator and mist eliminator. Even though venturi scrubbers are more expensive than other scrubbers, their collection efficiencies are higher for fine PM, which ranges 70–99% for particles larger than 1 μm. High gas velocities and turbulence in the venturi throat result in high collection efficiencies.

6. Orifice Scrubber

   The orifice scrubber, also called the impaction scrubber, allows the gas stream to flow over the surface of a pool of scrubbing liquid. The water droplets are entrained by the gas stream when it collides with the water surface. Then the waste gas flows upward, where it enters the orifice, a narrower opening. The orifice creates turbulence that can atomize the entrained droplets, where they capture the PM in the gas stream. A baffle series then removes the droplets from a liquid pool located below the scrubber. A recirculation pump is not required for circulating the scrubbing liquid, which is the main advantage of an orifice scrubber. The primary disadvantage is the difficulty of removing waste sludge. A sludge ejector is used to remove the waste sludge, operating like a conveyor belt.

4.2.1.1.2 Wet Scrubbers for Acidic Gases

For the purpose of scrubbing acidic gases, a scrubbing liquid is allowed to flow countercurrent to the waste gas flow, which entraps the acidic gases. A chemical aqueous solution with acid-neutralizing capability is used as the scrubbing liquid in the case of acidic gases with low water solubility. The discharging scrubbing liquid that absorbed the acidic gases is collected to enable the recovery of valuable substances though further treatment (Hegarty *et al.* 2000) .

The waste gas streams of production processes may include acidic gases such as:

- Hydrogen chloride
- Hydrogen fluoride
- Hydrogen bromide and hydrogen iodide
- Chlorine, bromine and fluorine
- Hydrogen cyanide
- Nitric acid

Sodium hydroxide (NaOH), potassium hydroxide (KOH) and sodium hypochlorite (NaOCl) aqueous solutions can be used as scrubbing liquid for most gases.

The following are some industrial applications of wet scrubbers in general:

- Industrial boilers
- Incinerators
- Metals processing
- Chemical production
- Asphalt production
- Fertilizer production

The advantages of wet scrubbers include:

- Wet scrubbers are smaller and more compact than dry scrubbers
- They have lower capital cost
- Operation and maintenance (O&M) costs are low
- They can be used to remove particle which:
    - Readily absorb water and tend to be sticky
    - Are combustible, corrosive and explosive
    - Are difficult to remove in their dry form
    - Contain high moisture content

The disadvantage of wet scrubbers include:

- High cost for increased collection efficiency due to an increased pressure drop across the control system
- Limited to lower waste gas flow rates and temperatures than dry scrubbers
- Waste generation in the form of a sludge that requires treatment and/or disposal
- Downstream corrosion or plume visibility problems can result unless the added moisture is removed from the gas stream

4.2.1.2 Dry Absorption (Dry Scrubbers)

In the dry absorption process, the separation of gaseous substances takes place through chemical or physical sorption agents that are put in contact with waste gas stream. The final products of the reaction accumulate in the form of dissolved or dry salts. The dry scrubbers control acid gas emissions such as SO_2, HCl, HF, etc. where they are primarily utilized in industrial boilers, municipal waste combustors, medical waste incinerators, and some refinery processes. Moreover, wet scrubbers function effectively as acid gas collectors.

The gaseous pollutants are bound to the surface of the introduced solid known as a sorbent through an absorptive gas–solid reaction. A fabric filter can be then used to separate the additives from the gas stream, which also allows for the removal of dust particles. This separation process is known as dedusting (Kroll and Williamson, 2012). Figure 4.3 shows the typical process of a dry scrubber.

There are two basic types of dry systems, which can be characterized based on the additive applied. They are:

1. Sodium-Based (Application of $NaHCO_3$)

In this type of dry system, sodium hydrogen carbonate ($NaHCO_3$) is used as the sorption material to remove acidic gas compounds. $NaHCO_3$ decomposes into sodium carbonate (Na_2CO_3), carbon dioxide (CO_2) and water (H_2O) at temperatures over 140°C. Na_2CO_3 further reacts with the acidic gases in order to convert them into harmless materials as they absorbed in the solids.

$$2NaHCO_3 \rightarrow Na_2CO_3 + CO_2 + H_2O$$

$$Na_2CO_3 + 2HCl \rightarrow 2NaCl + CO_2 + H_2O$$

This enables a good dedusting result for HCl and SO_2 at comparatively high temperatures and regardless of the flue gas humidity.

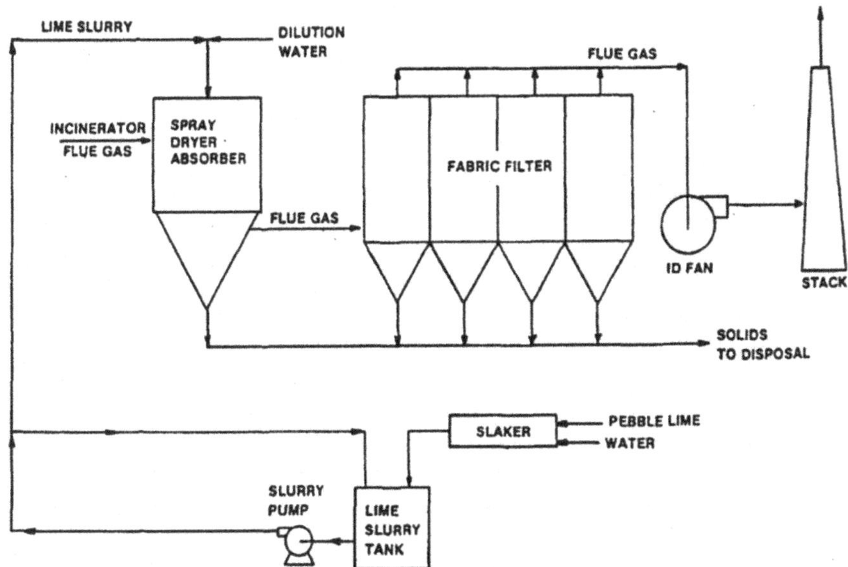

FIGURE 4.3 A dry scrubber.

2. Lime-Based (Application of $Ca(OH)_2$)

The separation of the pollutants is undertaken by allowing the waste gases to be come in contact with calcium hydroxide particles. Gas particles are adsorbed on the surface of calcium hydroxide through particular reactions. Some examples of the primary reactions are as follows:

$$SO_2 + Ca(OH)_2 \rightarrow CaSO_3 \cdot 1/2H_2O + 1/2H_2O$$

Reaction of lime with HCl in two steps:

$$Ca(OH)_2 + HCl \rightarrow Ca(OH)Cl + H_2O\,(l)$$

$$Ca(OH)Cl + HCl \rightarrow CaCl_2 + H_2O\,(l)$$

Conditioned dry sorption is also carried out using lime, wherein it separates pollutants via deposition on lime that is ameliorated by the hydrate shell that is formed around the lime particles. Accordingly, the separation is governed by the dissolution rate of the pollutants in aqueous solution, and both absorption and adsorption processes are run in parallel with the separation process.

Spray absorption is another method, which uses lime slurry. A spray dryer injects a lime and water suspension, referred to as lime slurry, into the flue gas stream. The slurry is created from quicklime (CaO) and water via an on-site slaking process. As the slurry is injected into the flue gas stream, water evaporates and only the solid lime particles remain in the flue gas. The liquid phase of this process has proven to be more efficient concerning HCl removal than the application of slaked lime (calcium hydroxide) in dry form.

In each method dry activated carbon or lignite coke can be dispersed into the flue gas stream to achieve better dedusting and to separate dioxins and furans from heavy metals (Karpf, 2015).

Advantages of dry absorption include:

- Low operative and maintenance requirements due to single-stage construction
- Low space required
- Comparatively inexpensive (The large quantity of additives that this type of process consumes is counterbalanced by the low purchase price of said additives)
- Energy consumption levels are low when compared to the energy levels required for wet flue gas cleaning systems

4.2.1.3 Electrostatic Precipitators (ESPs)

Electrostatic precipitators are among the most widely used particulate-collection devices that remove particles from a flowing gaseous stream (such as air). The device is operated with the force of an induced electrostatic charge. ESPs can be operated at high temperatures and high pressures with low power requirements. Thus, ESPs are often considered the ideal method for collecting small particles while obtaining a high efficiency. ESPs allow for the easy removal of fine particulate matter, such as dust and smoke, contained in an air stream, with minimum resistant to the gas flow. This categorizes it as an efficient filtration device. In the device, electrostatic attraction acts as the basic force that separates the particles from the gas. The particles are forced to pass through a corona, which gives an electrical charge to the particles. The corona is a specific region in which gaseous ions are subjected to an electrical field. Figure 4.4 shows an ESP unit. The electrical field forces the particles towards the walls (De Yuso *et al.*, 2013). The electrical field produces high voltage electrodes in the center of the flow lane. Industrial emission control serves following three purposes:

1. Recovery for economic reasons
2. Removal of abrasive dusts to reduce wear of fan component
3. Removal of objectionable matter from gases being discharged into the atmosphere

The followings are the major applications of electrostatic precipitators:

- Pulp and paper mills, non-ferrous metal industry, chemical industry, public buildings and areas
- Cement recovery furnaces, steel plants for cleaning blast furnace gas
- Removing tars from coke ovens, sulfuric acid (pyrite raw material), and phosphoric acid plants
- Petroleum industry for the recovery of catalysts and carbon black; thermal power plants

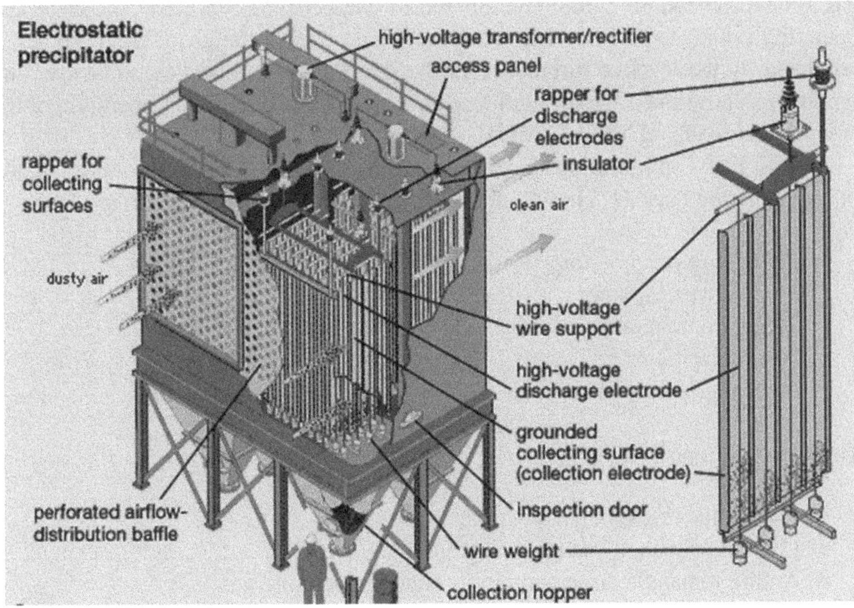

FIGURE 4.4 An electrostatic precipitator.

There are a few special requirements to be met to continue an ESP process. A high voltage source is required to generate the electrical field where discharge and collecting electrodes are essential. Appropriate gas inlet and outlet facilities should be located, alongside a suitable means for disposing collected material. Outer casing and a cleaning system should be fabricated for proper functioning of the system.

In an electrostatic precipitation process, first the corona is generated by the electrodes. As the potential difference between the wire and the electrode increases, it generates a voltage that can induce breakdown of the gas near the wire. This electrical breakdown or ion discharge, referred to as a corona, converts gas from its insulating state into its conducting state. Corona discharge can be either negative or positive depending on the polarity of the discharging electrode. The particle charging process then takes place due to the impact of a rain of negative ions generated from the corona process. This charging process takes place in the region between the boundary of the corona glow and the collection electrode.

Generally, two main mechanisms are involved in the particle charging processes, and these are influenced by the particle's size. The dominant force governs the particles that have diameters greater than 1 μm, while diffusion charging applies to particles smaller than 0.2 μm. The charged dust particles in the electric field move towards the collection electrodes. Those charged dust particles adhere and precipitate on the surface of the collection electrode. In liquid aerosols treatment, collected particles can be removed through coalescing and draining. Periodic impact or rapping processes can be used in the case of solid material. In solid material collection, a significantly thick layer of dust is collected such that it falls into the hopper or bin in a coherent masses to prevent excessive re-entrainment of the material into the gas system (Theodore, 2008).The amount of charged PM can affect the electrical operating point of the ESP. The movements and transportation of the particles is regulated by the level of turbulence in the gas. Particle properties, sneakage, and rapping re-entrainment greatly influence the total performance and operation of the system (Sudrajad and Yousuf, 2015).

Several ESP configurations have been developed either for special control action or for economic reasons. These types are:

- Plate–wire precipitator
- Flat plate precipitator
- Tubular precipitator
- Wet precipitator
- Two-stage precipitator

The advantages of ESP are:

- High collection efficiency
- Low maintenance and operating costs
- Ability to handle large volumes of high temperature gas
- Very low, or neglible, treatment time
- Easy cleaning

Some disadvantages of ESP include:

- High initial cost
- High space requirements
- Possibility of explosion hazards
- Possibility to produce poisonous gases

4.2.1.4 Fabric Filters

Fabric filtration is a physical separation process that utilizes a porous fabric medium. A fabric is a collection of fibers that are attached to each other, creating a permanent stable structure. Fibers are long, thin hair-like material with a length or diameter larger than 100 μm. When gas or liquid containing solids passes through the medium, the solids are retained. A combination of mechanical particle capture mechanisms such as impaction, straining, diffusion, and direct interception are used in fabric-filter operations. Straining is the process of removing particles larger than filter pores. Large particles are removed by impaction, while medium-sized particles are removed by direct interception and very small-sized particles are removed by diffusion. Impaction and direct interception possess a 99% collection rate of particles that are greater than 1 μm in aerodynamic diameter. Fabric filters can also collect very small particles, less than 1 μm in aerodynamic diameter.

Particles in a waste stream typically float along the gas stream, allowing the gas molecules to flow through the tiny pores in the fabric. These molecules create a continuous stream around the fiber in the filter. Retained solids can be periodically removed from the filter medium where the process operates in a batch or semi-continuous mode. Filtration systems may also be designed to operate in a continuous manner. As with other filtration techniques, an accumulating solid cake undertakes the majority of the filtration process. Thus, the formation of an initial layer of filter cake is essential at the beginning of the filtration operation (Air & Waste Management Association, 2007). The basic structure of fabric filters is as shown in Figure 4.5.

Fabric filtration effectively controls environmental pollutants in gaseous or liquid streams. It removes PM from gaseous emissions while filtering suspended solids in wastewater streams. Optimum use of a fabric-filter system would minimize problems with waste disposal. For example, in an air pollution controlling system it can be coupled with a scrubber to remove particles and/or gases from an emission source. The crystalline sludge that results from the reaction of the scrubber is filtered by a filter bag. A fabric filter may also be used to remove solids from water so that the water can be recycled; obviously, effluent slurry does not present a water pollution problem (Palazzo *et al.*, 1994). Fabric filters are considered efficient solid-removal devices with nearly 100% efficiency on occasions.

The efficiency of fabric filters depends on several factors (Wang *et al.*, 2004), including:

1. Dust Properties
   - Size: particles 0.1–1.0 μm in diameter may be more difficult to capture
   - Seepage characteristics: small, solid, spherical particles tend to escape
   - Inlet dust concentration: the deposit is likely to seal over sooner at high concentrations

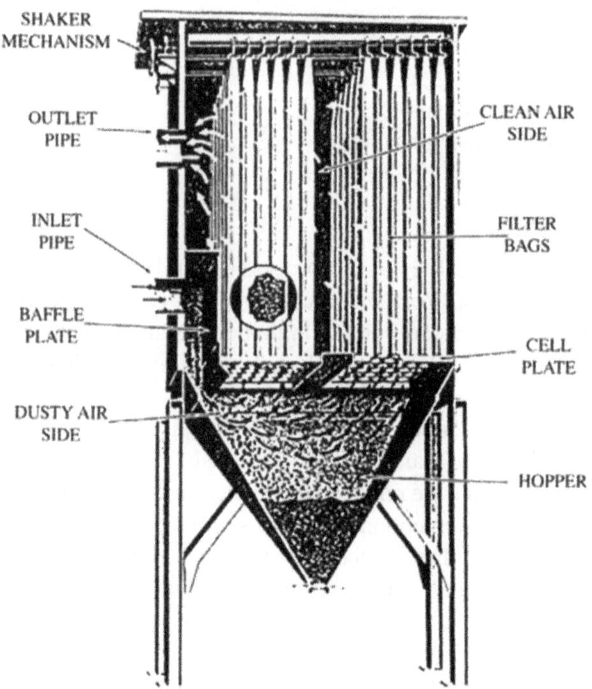

FIGURE 4.5 Typical structure of a fabric filter.

2. Fabric Properties
    - Surface depth: shallow surfaces form a sealant dust cake sooner than napped surfaces
    - Weave thickness: fabrics with high permeabilities, when clean, show lower efficiencies

The advantages of fabric filters include:

- High collection efficiencies for both coarse and fine (submicron) particulates
- Relative insensitivity to fluctuations in gas stream conditions
- Comparatively simple operation
- Simple maintenance
- Collection of particles with resistivities
- Collection of fly ash from low sulfur coals or fly ash containing high unburned carbon levels, which have high and low resistivities, respectively.

Some disadvantages of fabric filters include:

- The filter must remain dry when gases are being filtered – if moisture is present the PM tends to dissolve, blinding the filter cloth and resulting in a "mudded" filter fabric. Therefore, the fabric must be washed

- The filter must remain wet when liquids are being filtered – if the cake on the cloth is allowed to dry during liquid filtration, it will reduce the porosity of the cake. The result will be a partial blinding of the filter, which will then reduce the filtration rate

4.2.2 Liquid Waste Treatment

Production processes can result in various types of liquid waste effluents. These are:

- Liquid–liquid mixed waste (dissolved/emulsified)
- Liquid–liquid mixed waste (un-dissolved)
- Liquid–solid waste (dissolved)
- Liquid–solid waste (emulsified / dissolved)

Methods of liquid waste treatment are described below.

4.2.2.1 Filtration (Membrane Separation)

Various membrane separation processes exist and these can be classified by a number of criteria. These membrane separation processes are proven for use in purification, desalination, ion separation, metal recovery, and concentration processes. Membrane technology is useful in many industries, including pharmaceuticals, medical, chemical, and food processing. According to the process driving force, filtration methods can be classified as:

- Reverse osmosis
- Nanofiltration
- Ultrafiltration
- Microfiltration

The application of filtration methods varies according to pore sizes and driving force, as depicted by Figure 4.6. Table 4.3 shows the typical filtration methods and their applications.

4.2.2.2 Aerobic Treatment

Aerobic biological treatment is undertaken by aerobic microorganisms, mainly bacteria, in the presence of oxygen. They complete their metabolic activity using organic matter in the wastewater, producing the principle constituents of CO_2, NH_3, and H_2O. Among all other biological treatment methods, aerobic treatment is considered the most common and widely used process in the world. In oxygen-rich environments, the bacteria rapidly consume organic matter and convert it into carbon dioxide. But when the materials lack organic matter, the bacteria die and are subsequently utilized by the other, living bacteria as their food. This phase of the aerobic treatment process is called endogenous respiration. Significant solid reduction takes place in this phase. The capital cost of aerobic digestion is lower in comparison to anaerobic digestion, as aerobic digestion occurs within a comparatively short period.

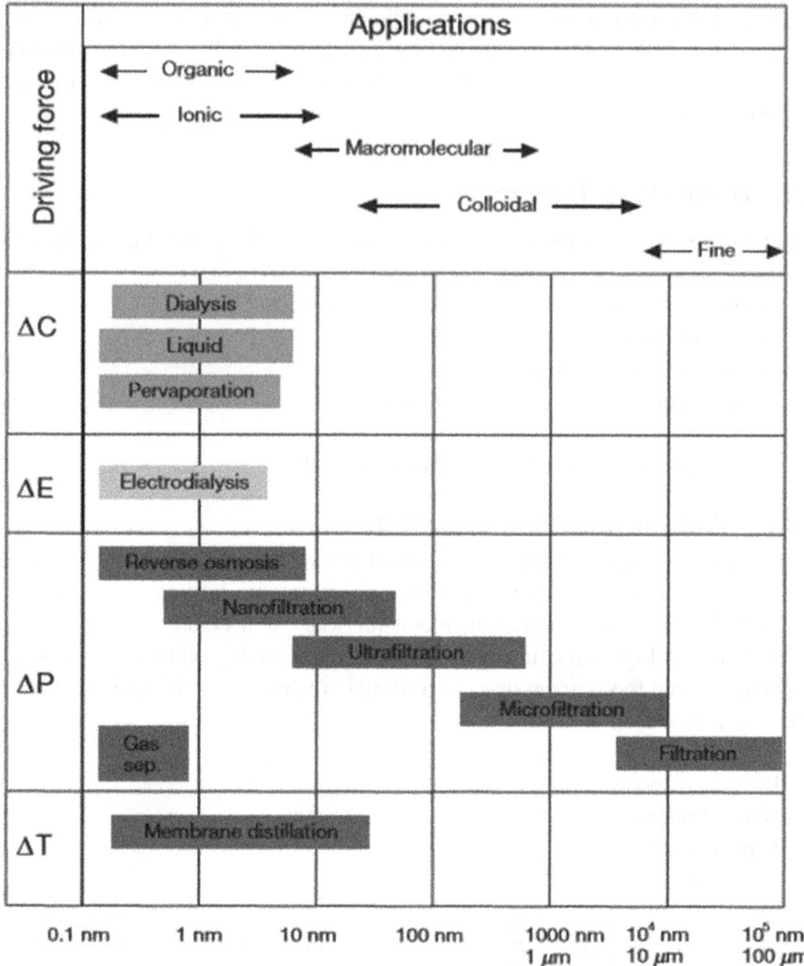

FIGURE 4.6 The application of filtration methods.

Nevertheless, the operating costs are considerably higher for aerobic digestion since the systems engender a higher energy cost because of the oxidization process.

There are several aerobic biological processes that differ according to their method of supplying oxygen and the rate of organic matter metabolization(Chan *et al.*, 2009).

4.2.2.2.1 Activated Sludge

In the activated sludge process, an aeration tank or basin acts as the dispersed-growth reactor, as shown in Figure 4.7. The reactor contains a suspension of the wastewater and microorganisms, known as the mixed liquor. The biological suspension is aerated by frequently mixing the contents of the aeration tank. Aeration devices commonly used for this purpose include submerged diffusers and mechanical surface

TABLE 4.3
Filtration Methods and Their Applications

Separation Process	Membrane Type	Separation Method	Range of Application
Microfiltration	Symmetric microporous membrane	Sieving mechanism as a function of pore size and adsorption	Sterile filtration clarification
Nanofiltration	Asymmetric skin type membrane	Solution diffusion mechanism	Separation of divalent ions from solutions
Ultrafiltration	Asymmetric microporous membrane	Sieving mechanism	Separation of macromolecular solutions
Reverse osmosis	Asymmetric skin type membrane	Solution diffusion mechanism	Separation of salts and micro-solutes from solutions
Electrodialysis	Cation and anion exchange membrane	Selective transport of ions or molecules according to electric charge	Desalting of ionic solutions

aerators. Submerged diffusers release compressed air while mechanical aerators introduce air by agitating the liquid surface. Hydraulic retention time in the aeration tanks is usually 3–8 hours but can be higher with high BOD_5 wastewaters. Following the principle of the activated sludge process, the methods of extended aeration and oxidation ditches have been commonly used.

After undertaking the aeration step, the sedimentation process allows the separation of microorganisms from the liquid; the clarified liquid is called secondary effluent. A portion of the biological sludge is recycled into the aeration basin to

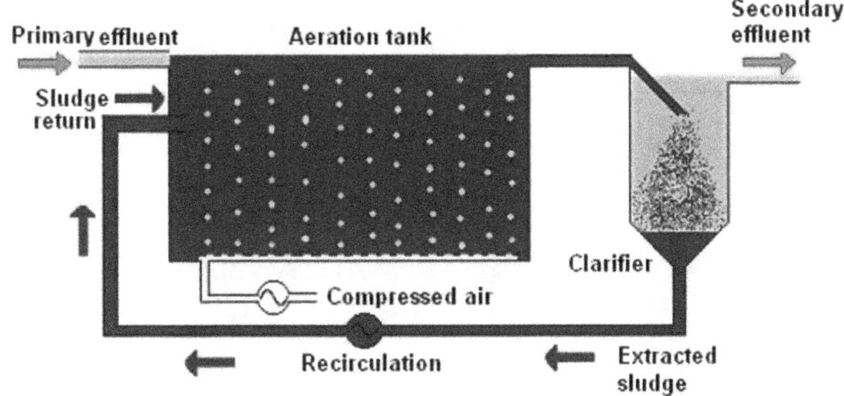

FIGURE 4.7 An activated sludge system.

maintain a high mixed-liquor suspended solids (MLSS) level. The remaining portion is removed from the process where it is sent for sludge processing. This processing will allow for the maintenance of a moderately constant concentration of microorganisms throughout the system.

4.2.2.2.2 Trickling Filters

A trickling filter is an attached-growth-type aerobic process where microorganisms are attached to a medium so as to remove organic matter from wastewater . Trickling filters may be circular with a rotary distributor, or stationary with a dosing chamber and a spray field. The following three components are mainly included in a trickling filter:

- Distribution system
- Filter media
- Under drain system

A trickling filter or biofilter consists of a basin or tower filled with support media where wastewater is applied occasionally or continuously. Figure 4.8 shows a trickling filter system. Crushed stones, wooden slats, rock, or plastic shapes can be used as the support media that microorganisms become attached to. The layer is then called a biological layer or fixed film. The ideal medium preferably has high specific surface area, high void space, low weight, biological inertness, chemical resistance, mechanical durability, and low cost. Some important characteristics of the medium include porosity, which measures the void spaces available for the passage of wastewater, air, and product gases, and specific surface area, which refers to the amount of the surface area of the media that is available for the biofilm's growth. Size of the medium ranges 50–100 mm, with a specific surface area in the range of 50–65 m^2/m^3 with porosities of 40–50 % (Davis and Cornwell, 2008).

Wastewater influent is normally supplied from the top of the trickling filter under a hydraulic head of about 1.0 m through the jet action of nozzles. When the flow is intermittent, sufficient air circulation is supplied through the pores between the dosing. A distributor arm aids in the continuous and uniform distribution of wastewater over the film.

As the wastewater moves through the filter, the organic matter is adsorbed onto the film and microorganisms metabolize the organic matter in the wastewater diffuse

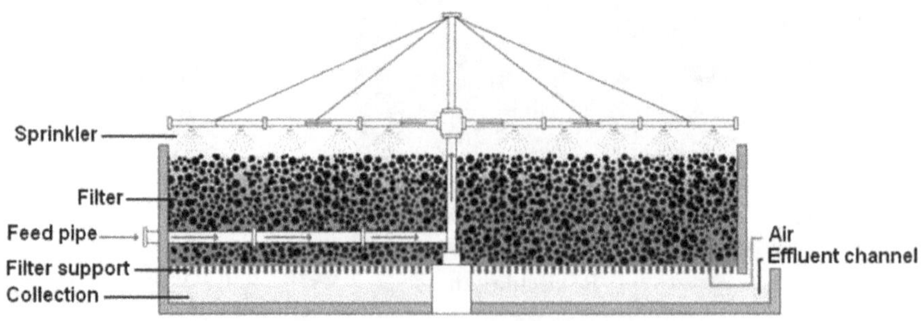

FIGURE 4.8 A trickling filter.

(Henze *et al.*, 2001). Natural air flow supplies oxygen to the film while blowers are used to supply forced air. A forced air supply is rarely required as natural air flow can be regulated from an up- or downward direction depending on the relative temperatures of the wastewater and ambient air. The thickness of the biofilm increases as new organisms grow. With the continuous growth of biological film, a portion of the slime layer falls off the filter as the microorganisms near the surface lose their ability to cling to the medium. This process is called sloughing. Sloughed material from the media is separated in a secondary clarifier and discharged for sludge processing. Facultative bacteria such as Achromobacter, Flavobacterium, Pseudomonas, and alcaligenes are mainly used in the process to decompose the organic material in the wastewater. Generally, in the lower reaches of the filter, nitrifying bacteria can be present (Adams *et al.*, 1997).

The main factors that affect the operation of trickling filters are as follows:

- Organic loading
  - A high organic loading rate results in a rapid growth of biomass
  - Excessive growth may result in the plugging of pores and a subsequent flooding of portions of the medium
- Hydraulic flow rates
  - Increasing the hydraulic loading rate increases sloughing and helps to keep the bed open
- Relative temperature of wastewater and ambient air
  - Cool water absorbs heat from the air, and the cooled air falls towards the bottom of the filter, concurrent to the water
  - Warm water heats the air, causing it to rise through the underdrain and up through the medium
  - At temperature differentials of less than 3–40°C, relatively little air movement results, and stagnant conditions prevent good ventilation
  - Extreme cold may result in icing and destruction of the biofilms

4.2.2.2.3 Rotating Biological Contactors

Rotating biological contactors (RBCs) are fixed-film reactors. They are similar to biofilms, where microorganisms are also attached to support media. In RBCs, the support media are slowly rotating discs that are partially submerged in flowing wastewater in the reactor. Oxygen is supplied to the attached biofilm from the air when the film is out of the water, and from the liquid when submerged, since oxygen is transferred to the wastewater by surface turbulence created by the discs' rotation. Sloughed pieces of biofilm are removed in the same manner described for biofilters (Metcalf and Eddy, 2003).

The advantages of trickling filters are:

- Low energy requirements: Trickling filters do not typically require additional energy-consuming equipment such as aeration blowers
- Easy to dewater waste sludge: Sludge from trickling filters is primarily composed of the sloughed off biological slime layer that tends to settle and dewater more easily than waste activated sludge from conventional activated sludge plants

- Low maintenance cost: Trickling filters do not consist of many moving parts, which leads to minimum maintenance requirements
- Consistent effluent quality: Typically, the trickling filter technique that performs extremely reliably at low or consistent loadings
- Resistant to toxins and shock loads: Trickling filters can handle and recover from shock loads since they are not a complete mix system. A toxic "slug" will only affect the portion of the filter that it is sprayed on, thus allowing the remaining portion to continue its operations in the normal manner. Shock loads that are introduced to the unit will be diluted due to the recirculation of filter effluent back through the media
- Ease of operation: Trickling filters do not require a high level of sophisticated operation in order to provide a reasonable effluent

The disadvantages of trickling filters are:

- Odors and unfavorable organisms: Obnoxious odors can generate due to the anaerobic decomposition that is caused by excessive organic loading or inadequate ventilation of the filter media. In order to reduce the problem in covered filters, a forced-air ventilation system and odor control of the exhaust is usually provided. Through proper maintenance and checking, atmospheric hazards can be eliminated. Filter flies (Psychoda) and other insects can thrive around the filters as a result of low maintenance or inadequate moisture in the filter media
- Potential of clogging the media: Portions of the filter media may be clogged due to sloughing off of the media slime layer. This can lead to inefficient treatment removal and poor effluent quality. Seasonal sloughing will be apparent from the increase in secondary sludge production
- Freezing in cold weather conditions: In temperate regions, the extreme cold weather conditions of winter months can cause distributor arm orifices or spray nozzles to freeze, resulting in low hydraulic loading onto the filters
- Low potential for adjustments: Trickling filters do not have features that allow them to be quickly adjusted for a rapid increase in loading. In addition, trickling filters cannot be fine-tuned to achieve a high level of treatment
- Pumping Costs: It may be necessary to pump the wastewater to a higher elevation so that the flow exit through the distributor. Additionally, recirculation of wastewater may be necessary to achieve sufficient wetting of the media

When considering the aerobic digestion process of waste treatment as a whole, there are some advantages as well as disadvantages. These are simply stated below.

The advantages of aerobic digestion are as follows:

- Aerobic bacteria are very efficient in digesting and breaking down waste materials
- Aerobic treatment typically results in better effluent quality in comparison to the effluent quality obtained through the anaerobic digestion process

- The aerobic digestion process releases a significant amount of energy
- A portion of the sludge is utilized by the microorganisms for synthesis, growth, and development of new microorganism communities

The major disadvantages of the aerobic digestion process are that:

- Substantially great cost is associated with supplying power for the aeration of the system by providing the required oxygen
- A digested sludge is produced with poor mechanical dewatering characteristics
- The digestion process is restricted and greatly depends on some external and internal dominating factors such as temperature, location, and type of tank material
- Useful by-products (such as methane, which is recovered through the anaerobic digestion process) are not obtained
- More sludge is produced for disposal

4.2.2.3 Anaerobic Treatment

Anaerobic treatment is a well-established process utilized for treating waste and wastewater and, recently, in biofuel production. Anaerobic digestion is the process of breaking down organic materials into simpler compounds in the absence of oxygen. Anaerobic treatment is undertaken with anaerobic bacteria (biomass), as these consume the biodegradable organic pollutants in their metabolic activities. There are two main purposes of anaerobic digestion: it can be used to treat and reduce biodegradable wastes, and to produce salable products such as heat, electricity, and soil amendments. Before subjecting the materials for digestion process, pre-treatment should be undertaken. Pre-treatment methods vary according to the type of waste material or the feedstock. The pre-treatment process is carried out with the purpose of removing undesirable materials such as large unwanted and inert materials (plastic, glass), mixing different waste materials, and adding water. The pre-treatment obtains digestate of a better quality, while enhancing the efficiency of the digesting process and ensuring that the process is free from failures.

The anaerobic microorganisms convert organic materials into carbon dioxide, methane, and a small amount of biosolids in the absence of oxygen. Methane, which is considered a biogas, is an energy rich compound that can be used as boiler feed and/or as combined heat and power (CHP) to produce "green" electricity and heat. The anaerobic sludge contains various groups of microorganisms that work together in converting organic material into biogas through the processes of hydrolysis and acidification. The sludge, which is a valuable by-product of the process, can be used as nutrient-rich manure for crops (Lettinga *et al.*, 2001). The typical structure of an anaerobic digestion system is as conveyed in Figure 4.9.

Although there are numerous kinds of digesters, the biological basis is the same for all. Anaerobic digesters are built systems that consciously harness the natural process of anaerobic digestion. These anaerobic digesters are fabricated in a manner that can minimize odors and vector attraction, and reduce pathogens and waste volumes, while inducing gas production, and liquid and solid digestate production.

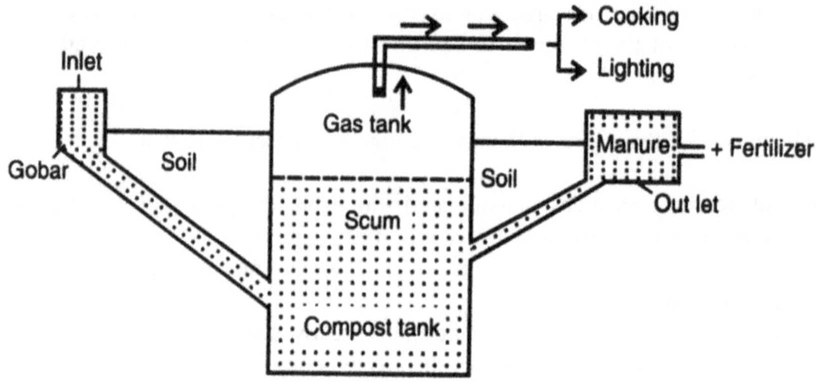

FIGURE 4.9 Anaerobic treatment.

Different materials such as brick, steel, concrete, and plastic can be used to construct the digesters. The shape of the digester may vary from silos and ponds to basins and troughs, where they can be located either on the surface or underground. The components required for the operation of anaerobic digestion system are as follows:

- Premixing tank or area
- Digestion vessel
- Biogas collection and storage point
- Effluent collection and distributing system

Several anaerobic digestion systems and processes have been developed and are categorized for various waste types:

1. Upflow anaerobic sludge blanket (UASB) for slurry
2. Dry batch and dry continuous processes for high solid substrates
3. Sequential batch process for organic fraction of municipal solid waste
4. Two-phase anaerobic digestion system for organic fraction of municipal solid waste

Anaerobic digestion can be performed as a batch process or a continuous process. In batch processes, biomass is added to the reactor at the start of the process and sealed within the digester throughout the duration of the process. In continuous digestion processes, organic matter is constantly added or added in stages to the reactor, with the end products removed periodically. It ensures the constant production of biogas (Korres *et al.*, 2010).

There are some factors that determine the type of anaerobic digester to use. Depending on the following factors or a combination of those factors, anaerobic digester types can vary:

- Dry solid content of substrate: Dry or wet anaerobic digester
- Type of substrate: High solid or low solid anaerobic digester

- Mode of substrate feeding: Batch or continuous anaerobic digester
- Operating temperature of the process: Mesophilic or thermophilic anaerobic digester
- Complexity of the digestion process: Single-stage or multistage anaerobic digester

A significant reduction in large organic matter quantities occurs during the process of anaerobic digestion. Anaerobic digestion reduces the amount of organic waste that would otherwise be incinerated or dumped in the sea or in landfills. Anaerobic digestion is particularly useful to treat waste effluents emitted from industrial and domestic processes. It can range from paper waste, food leftovers, grass clippings and botanical wastes, to sewage and organic farm wastes. Woody wastes are not recommended for use as the substrate in the process since they contain high amount of lignin, which is hard to digest through anaerobic digestion. When the digester is fed with more than one substrate the process is called co-digestion; animal manure or slurry and other organic wastes are typically a better combination. The effluent waste that is emitted by wet scrubbers can be used for biogas production through the process of anaerobic treatment.

Methane is a promising energy component produced in anaerobic digestion. Apart from this valuable energy source, biogas also comprises some non-methane components such as hydrogen sulfide, carbon dioxide, and water vapor. These materials tend to inhibit methane production. Except for water vapor, they are generally considered to be harmful to humans and/or the environment. Thus, the produced biogas has to be appropriately cleaned using suitable scrubbing and separation mechanisms.

In addition, methane is a difficult to detect gas since it is odorless and colorless. Therefore, it has a considerable potential to pose a huge threat. Methane is a highly explosive gas when it comes in contact with atmospheric air with a 6–15% proportion. To eliminate these issues, it is recommended that buildings are kept well ventilated. Motors, wiring, and other lights should be explosion proof; fire alarms and gas detection systems can be used as preventive methods, where flame arrestors have to be used on gas lines.

Causes of digester failures in anaerobic digestion include:

- Unbalanced microbiological growth
- Acid formers out-producing methane formers
- Overproduction of acids

Process failure indicators include:

- Increasing volatile acid (VA) concentration
- Dropping alkalinity (Alk)
- Increasing VA/Alk ratio
- Falling pH
- Dropping gas production rate
- Increasing percentage of CO_2

Some advantages of anaerobic treatment are:

- The formation of biogas, which has a high energetic value: Energy generated through the biogas in anaerobic digestion can reduce the demand for fossil fuel
- Higher volumetric loads (COD) are achieved compared to aerobic wastewater purification (typically 5–10 times higher)
- Very little sludge production (4–5 times lower than in an aerobic system)
- Less greenhouse gas emissions: A properly managed anaerobic digestion system aims to maximize methane production without releasing any gas into the atmosphere. This reduces the overall emission of gases that contribute to global warming and climate change
- Low cost and environmentally sound waste treatment method: Anaerobic digestion can be used as an integrated management system to reduce the potential for soil and water pollution. If the products are used on site, this contributes to the economic viability of a particular farm by keeping the costs and benefits within the farm
- Improved fertilization efficiency: The digestate of the process can be used as a source of fertilizer that is organic in its means. This will reduce the extra energy demand for fertilizer production and decrease synthetic fuel usage in fertilizer manufacturing

The disadvantages of anaerobic treatment can be outlined as follows:

- Potential for incomplete breakdown of organic compounds
- The digestion process emits considerable amounts of effluents into the air, soil, and water, which requires frequent treatment to avoid damage to human and environment health
- The water produced during the process may be contaminated with nitrates and other chemicals; it is recommended that the water be processed and purified before being released into the environment
- Potential for odor problems
- The most efficient purification occurs in the mesophilic range, whereby the influent must be heated in most cases
- Less adaptability with regards to toxicity and inhibition

With the integration of the above treatment techniques, GP strategies can be enhanced by separately sorting the waste sources, and through the extent of required treatment facilities. For example, the wastewater treatment layouts in Figure 4.10 and Figure 4.11 can be suggested for low polluted wastewater and highly polluted wastewater effluents that result from production processes, respectively.

4.2.3 Solid Waste Treatment

Solid waste management is typically defined as being all the processes, activities, or programs that aim to reduce the pollution caused by solid wastes.

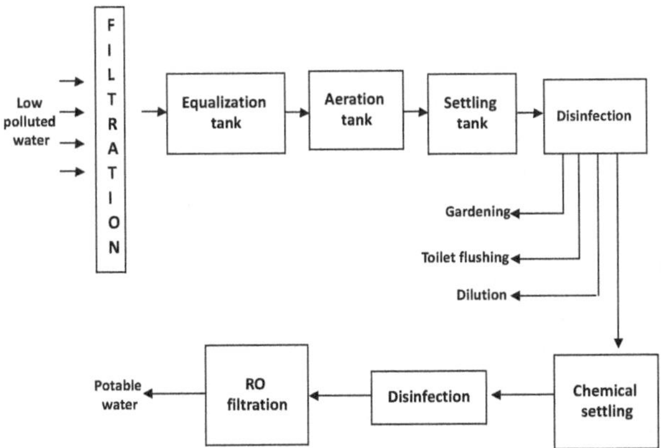

FIGURE 4.10 Layout of low polluted wastewater treatment.

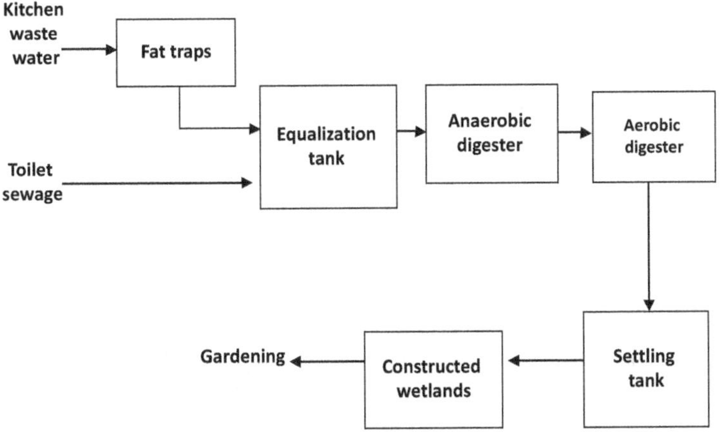

FIGURE 4.11 Layout of highly polluted wastewater treatment.

The major components of solid wastes are:

- Residual solid wastes/sludge from effluent treatment plants (e.g. from thickeners, filter presses, or sludge drying beds)
- Residual solid wastes from air pollution control equipment (e.g. particulate from bag filters)
- Direct process solid waste (e.g. tank bottoms, stills, etc.)
- Non-process solid wastes (e.g. unused raw materials, containers, packaging material, etc.)

The major solid waste treatment techniques are listed below.

4.2.3.1 Thermal Treatment

Thermal treatment is the processes of using heat to treat waste. Some commonly utilized thermal treatment processes are discussed below.

4.2.3.1.1 Incineration

Incineration is the most common thermal treatment process, and involves the simple combustion of waste in the presence of oxygen. Incineration usually refers to the combustion of unprepared (raw or residual) municipal solid wastes. A sufficient quantity of oxygen should be supplied to allow the full oxidization of the fuel. Generally, the temperatures of the combustion flame in the incineration plant exceed the 850°C threshold at which point the waste is converted into carbon dioxide and water vapor. Noncombustible materials such as metals and glass remain at the bottom as ash, a solid which may also contain a little amount of residual carbon.

Municipal solid waste incineration systems are found at the most advanced level of the waste disposal/treatment hierarchy as mechanical treatment methods (for example, composting and incineration). In waste incineration, additional environmental control is required at each level and therefore disposal costs increase substantially. Introducing the mechanical treatment of municipal solid waste requires a significant conversion of technology as well as costs.

This method may be used as a means of recovering energy for use in heating or the supply of electricity. Incineration is emphasized for its inherent advantage of reducing the volume of waste that would be otherwise moved into sanitary landfills (Assamoi and Lawryshyn, 2012). Hence, waste incineration plants are generally introduced in areas where sanitary landfills are in conflict with other interests such as city development, agriculture, and tourism.

Generally, construction, demolition, and street sweeping wastes are not recommended for use in incineration. Street sweepings are mainly comprised of dust and soil together with varying amounts of paper, metal, and other litter from streets. When considering lower-income countries, street sweepings may also include drain and domestic waste dumped along the roads, or plant remains and animal manure. Construction and demolition wastes may mostly comprise of soil, stone, brick, concrete and ceramic materials, wood, and packaging materials; although the composition of this waste depends on the type of building materials.

Different types of thermal treatments are applied to different types of wastes, with each method having its own merits and demerits depending upon the type of wastes to be treated. The following are some types of incineration:

- Grate incinerators
- Rotary kilns
- Fluidized beds

Although several types of incineration technologies are currently available, mass burning incineration with a movable grate is the most widely used. Mass burning incineration with rotary kilns is used to a lesser extent. Fluidized bed incineration is still at the experimental stage, while the mass burning technology with a movable grate has been successfully applied for decades. Mass burning technology has developed to comply with the latest technical advancements in parallel to environmental standards. Mass burning incineration can generally handle municipal waste without pre-treatment as it can work with the appropriate waste types as they are received.

Mass burning technologies are generally adequate for the large-scale incineration of both mixed and source-separated municipal and industrial waste types. Compared to movable grates, rotary kiln incineration plants have a smaller capacity and are mostly used for special types of waste unsuitable for burning on a grate, such as various types of hazardous, liquid, and infectious waste.

Municipal solid waste can be incinerated in several combustion systems including travelling grates, rotary kilns, and fluidized beds. Fluidized bed technology requires municipal solid wastes to be of a certain particle size range, which usually requires some degree of pre-treatment and/or the selective collection of the waste.

The incineration of sewage sludge mostly takes place in rotary kilns, multiple hearths, or fluidized bed incinerators. Co-combustion in grate-firing systems, coal combustion plants, and industrial processes is also applied. Sewage sludge often has a high water content and therefore usually requires drying, or the addition of supplementary fuels to ensure stable and efficient combustion.

For the incineration of hazardous and medical waste, rotary kilns are most commonly used, but grate incinerators (including co-firing with other wastes) are also sometimes applied to solid wastes; fluidized bed incinerators are used for some pre-treated materials. Static furnaces are also widely applied at the on-site facilities of chemical plants.

Other processes have been developed based on the de-coupling of phases that also take place in an incinerator: the drying, volatilization, pyrolysis, carbonization, and oxidation of waste.

The advantages of incineration include:

- Reducing waste volume and weight, with a nearly 90% volume reduction and 75% weight reduction
- Immediate waste reduction that does not require long-time residency
- Potential to be carried out onsite, reducing transportation costs
- Air discharges can be controlled
- Ash residue is usually non-putrescible, sterile, and inert
- No large disposal area for the residuals is required
- Cost can be compensated by heat recovery

The disadvantages of incineration include:

- Potential for contaminating air
- Greater cost for pollution gas cleanup
- Potential for releasing volatile organic compounds, carbon monoxide, ash, etc.
- Health risk for people

4.2.3.1.2 Pyrolysis and Gasification

Pyrolysis and gasification are considered to be similar processes. Both processes decompose organic waste by exposing it to high temperatures and low amounts of oxygen. The main difference is that gasification uses a low-oxygen environment

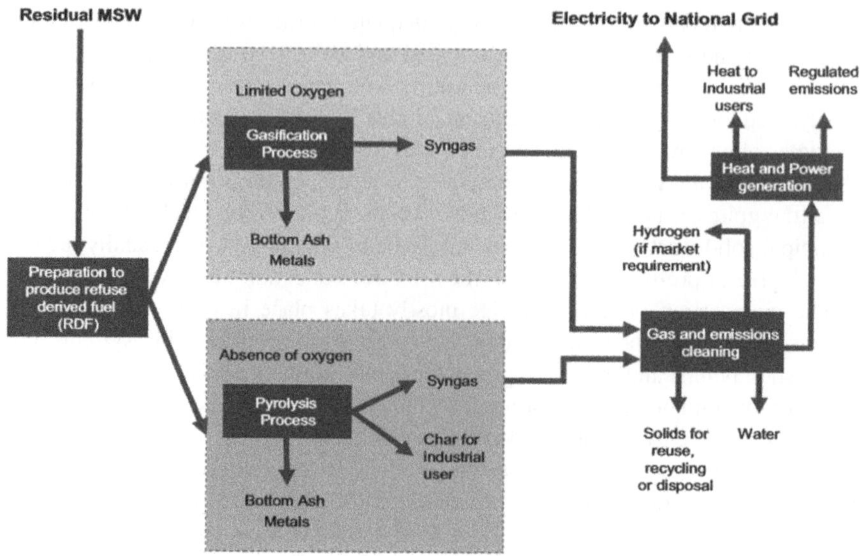

FIGURE 4.12 Pyrolysis and gasification.

while pyrolysis uses no oxygen. These techniques use heat and an oxygen starved environment to convert biomass into other forms, as shown in Figure 4.12.

The pyrolysis process requires an external heat source to maintain the adequate temperature. Moreover, temperatures of about 300°C to 850°C are used during the pyrolysis of materials such as municipal solid wastes. The required temperatures are relatively low. The pyrolysis process produces solid residue and a synthetic gas that is typically called syngas. The solid residue, sometimes known as char, is a mixture of non-combustible materials and carbon. The syngas is a gas mixture that involves combustible constituents of carbon monoxide, hydrogen, methane, and a broad range of other volatile organic compounds. A proportion of these can be condensed to produce oils, waxes, and tars. Typically, the syngas has a net calorific value (NCV) of 10–20 MJ/Nm^3. As per requirements, syngas is allowed to cool and the condensable fraction, which has the potential to utilize as a liquid fuel, can be collected.

In the gasification process the partial oxidation of substances occurs by adding oxygen amounts that are insufficient to allow the fuel to be completely oxidized, meaning full combustion does not take place. The required temperatures are typically above 650°C. The process is largely exothermic although some heat may be required to initialize and continue the gasification process. The main product is a syngas, which is comprised of carbon monoxide, hydrogen and methane. Generally, the generated gas of gasification may have an (NCV) of 4–10 MJ/Nm^3. Apart from these main products, gasification also produces a solid residue from noncombustible materials (ash), which may contain a moderately low level of carbon. For reference, the calorific value of syngas from pyrolysis and gasification is far lower than natural gas, which has a net calorific value of around 38 MJ/Nm^3.

A mixture of combustible and non-combustible gases as well as pyroligneous liquid is produced by these processes. The composition of the syngas and solid residue will depend on process conditions such as:

- Oxygen level
- Operating temperature
- Heating rate
- Residence time in the reactor

Apart from these factors, there are some others that may also influence the process, such as the direction of gas flow (whether the gas flow is horizontal or vertical). All of these products have a high heat value and can be utilized. The most commonly used configuration in producing energy is to burn the syngas in a boiler to generate steam. The generated steam can then be used to produce electricity by passing it through a steam turbine. Taking into consideration the demand for the process, heat generation can be done. The production of electricity, and the overall energy efficiency of the system, can be enhanced accordingly.

The syngas can also be used to fuel a dedicated gas engine that may also have the potential to be used in vehicles once it is reformed to produce hydrogen. It is predicted that the initial market for hydrogen would be public transport fleets using fuel-cell vehicles. In addition to using the syngas to produce energy for the above sources, it could also be used as a chemical feedstock. Gasification is advantageous since it allows for the incineration of waste with energy recovery, and without the air pollution that is characteristic of other incineration methods.

Both the pyrolysis and gasification processes are focused on treating the biodegradable materials that are present in municipal solid waste such as paper, cardboard, putrescible waste, green waste, wood, and plastics. Thus, it is considered a common method for removing non-combustible materials and recyclables, which typically consists of metals and glass. This treatment occurs prior to the primary treatment reactor stage. Moreover, depending on the utilized technology, the feed material may require pre-processing in order to remove excess moisture or to reduce size via a mechanism of shredding.

In terms of technology, rigorous evaluation is required in order to reduce any operational risks when processing the anticipated feedstock. With urbanization and other technological advancements, waste composition is likely to change. Thus, the pyrolysis and gasification processes should be robust, flexible, and adjustable enough to treat a considerable range of calorific values and compositions of waste feedstock.

4.2.3.2 Sanitary Landfills

Sanitary landfills are designed to reduce or eliminate the threats of waste disposal to public health and the environment by utilizing the principles of engineering. Using these principles, solid waste is confined to a minimum practical area, with the aim of reducing it to the smallest practical volume. The solid wastes are covered with layers of earth and geotextiles. Monitoring and regulation must be carried out at the conclusion of daily operations, or at more frequent intervals, in order to reduce the potential for risks or threats (Wall and Zeiss, 1995).

Usually, landfills are constructed in areas where land features act as natural buffers between the landfill and the environment. For example, the area may be comprised of clay soil, which is fairly impermeable due to its tightly packed particles, or the area may be characterized by a low water table and an absence of surface water bodies, thus preventing the threat of water contamination (Renou *et al.*, 2008). Normally, the selection of a site is commenced by searching for appropriate, conveniently located land with a low value; often, there is a focus on wastelands.

Available soil data and geological characteristics of the proposed location are important to determine using on-site field tests or with the aid of existing reliable survey data for the area. Geological investigations are one of the most important studies required in selecting a site. Using the acquired data, the planners and engineers are able to determine whether the existing on-site soil is suitable to utilize as the covering material of the landfill or whether it has to be obtained elsewhere. Moreover, these investigations will also aid in evaluating some other important geological factors that may ultimately lead to ease of excavation, water pollution, or lateral gas movement. A combination of approximately 50%clay-silt and 50%sand is considered an ideal soil for cover material. Thus, a sandy-loam mixture of soil is porous, well-compacted, and not susceptible to cracking upon drying. Clay soils are always susceptible to cracking when they become dry and those soils are difficult to handle when they become wet in the presence of water. This allows rodents and other insects to access the landfill pit through its cover.

The exposure of trash to wind and rain is minimized by dividing the landfill into a series of individual cells and filling only a few cells of the site at any one time. The desired elevation of the completed site, the site's characteristics, and good engineering practice are all considered as governing factors of the total landfill depth. The generally recommended maximum depth for a single cell is 8 feet, a depth which is typically accepted as avoiding the excessive settlement and surface cracking that results in deeper cells. Although fills which are shallower than 8 feet allow the site to be re-used earlier, they do not generally make maximum use of the available land.

Daily waste is spread and compacted to reduce volume. An adequate compaction of the waste is very important in producing a successful sanitary landfill. If the waste is not compacted well it may cause excessive settlement and uneven spreading. Solid waste should be placed at the top or bottom of the working face, spread in layers about 2 feet thick, and compacted. To avoid odors and to keep out pests, a cover is then applied. The landfill is sealed with an impermeable layer at the end, when it has reached its maximum capacity. Typically, the sealing material may be composed of soil (Pohland and Kim, 1999).

The solid wastes buried inside the soil are considered, under some geological conditions, a significant source of chemical and bacteriological pollution in ground water and surface water sources. At the site selection and planning stages, prior to developing the landfill site, water pollution potential should be thoroughly investigated. The soil type and topography can induce a pollution hazard that can increase when the disposal site is located in sand or gravel soil, both of which facilitate the rapid penetration of the leachate into the groundwater where it mixes with the water in the wells that are located nearby (Liu *et al.*, 2006).

Therefore, normally landfills are recommended not to be positioned on a site where ground water or surface water could intercept the dumped waste. In addition to strategic placement, preventive and protective measures should be incorporated into the landfill design in order to minimize and prevent potential pollutions resulting from sanitary landfill operations. The bottom and sides of landfills are lined with layers of clay or plastic to keep the liquid waste, known as leachate, from escaping into the soil. The leachate is collected and pumped to the surface for treatment. Leachate collection and treatment can be undertaken using the following steps:

- Establish leachate pipes at the base liner of the landfill
- Position a steady decline to a low point outside the landfill body
- Collect the leachate in a circular pipeline around the landfill
- Utilize intermediate storage in a leachate pond
- Utilize a leachate treatment plant or dispose to a wastewater treatment plant
- Introduce the it into water, a river, or a sewer

Leachate collection and treatment can be carried out through a multi-stage leachate treatment plant, established on-site. The multi-stage leachate treatment plant typically consists of following features:

- Biological Pre-Treatment
    - In biological pre-treatment, nitrification and de-nitrification of leachate occurs (tanks or basins can be used for the process)
- Membrane Filtration
    - The leachate is subjected to 7 bar excess pressure, which facilitates the retention of bacteria
- Chemical Wet-Oxidation
    - In this step, ozone treatment occurs, which can result in the cracking of long-chain hydrocarbons
- Biological Retreatment
- Percolation or a trickling filter is used for further treating the leachate

Groundwater quality is frequently monitored using wells that are dug in the surrounding area of the landfill. Figure 4.13 shows the basic parts of a sanitary landfill. Some common ground water pollution preventive measures are listed below:

1. Positioning the landfill site at a considerably safe distance from streams, lakes, wells, and other water sources
2. Avoiding site location above the kind of sub-surface stratification that will lead the leachate from the landfill to water sources (e.g. fractured limestone)
3. Using an earth cover that is nearly impervious (typically geotextiles)
4. Providing suitable drainage trenches to carry the surface water away from the site
5. Diking and de-watering, or draining and filling

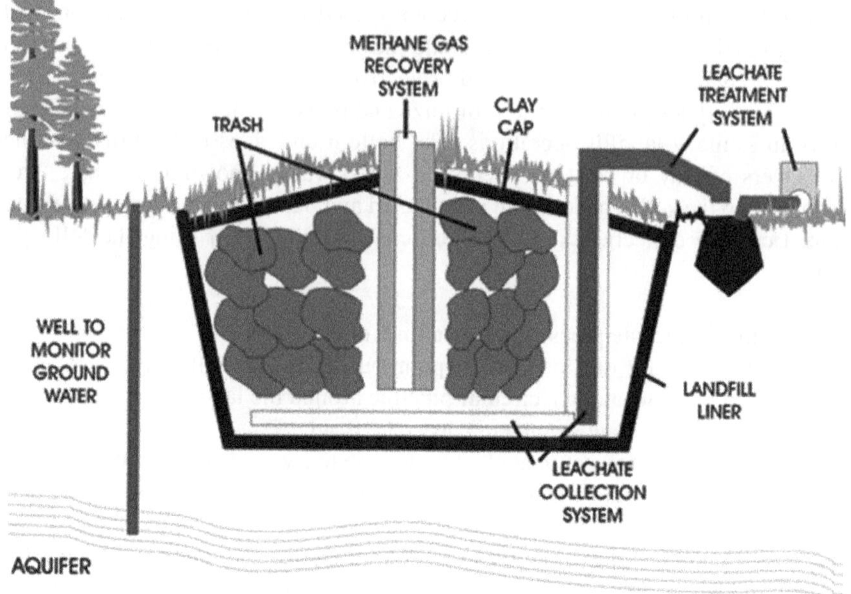

FIGURE 4.13 A sanitary landfill.

The surface of the landfill should be carefully graded to permit drainage and prevent the ponding of surface water. Standing water may permit mosquito breeding and may interfere with the operation of the landfill. Precautions must be taken to prevent runoff water from eroding the cover material and exposing the waste.

A permanent perimeter fence and plantings are essential for a land filling site if the site is not supported by a natural screening or afforded with trees or topography. This is considered a critical requirement for extending the useful life of the site over a period of time, for instance more than five years.

The confinement of blowing papers and light solid waste material to the landfill site is best accomplished by use of light, movable fences, such as snow fences. Such fencing should be located around the landfill operation to protect surrounding areas (Renou *et al.*, 2008).

The eventual use of the finished landfill should be pre-determined before the landfill operation commences, as this directs towards effective planning and operation. This should be done prior to the beginning of landfill operations as it can provide an aim and direction to the whole operation. Completed landfills have been most often used for recreational purposes such as parks, playgrounds, or golf courses. Parking lots, airports, storage areas, and botanical gardens can also be built on the completed landfill sites. Building construction is not recommended on the sites, since settling and gas problems may threaten the activities. For the purpose of sealing the landfill, the following materials can be used:

- Natural mineral material with a very low permeability, high clay content and a high content of swelling material (e.g. smectites)
- Sand–bentonite mixtures (swelling material)

- Geomembrane (HDPE), a thick, dense material that can resist mechanical properties such as UV, chemical, biological, and cracking stresses

Sanitary landfills are used for recovering energy through the anaerobic decomposition of waste. Due to natural decomposition, landfills produce carbon dioxide, methane, and traces of other gases that are known as landfill gases. Methane can be used as an energy source to produce heat and electricity. To collect the producing methane gas, landfills are fitted with landfill gas collection (LFG) systems. The process of generating gas is very slow, and for the energy recovery system to be successful there needs to be large volumes of wastes.

The advantages of landfills are:

- The least environmental impacts
- Low health risk
- Records kept can be a good source of information for future use in waste management

Some disadvantages of landfills are:

- High cost of establishing
- Potential for ground water contamination

4.2.3.3 Composting

Composting is a controlled process driven by microorganisms and small invertebrates capable of the aerobic decomposition of organic matter. Put simply, composting is the natural process of "rotting" that breaks down fractions of organic waste into stable substances. Compost preparation is considered one of the most effective and successful processes for recycling organic wastes. In composting, the natural process of organic matter decomposition is intentionally used to derive a natural fertilizer that is a valuable product for use in agriculture. These compost-producing microorganisms may be assisted by many other larger organisms. During composting, these microbes produce carbon dioxide, heat, and water while they break down the materials in the pile (Palm *et al.*, 2001).

The end product "compost", which results from the process of bacteria, fungi, and other microbes, is a nutrient-rich, dark, crumbly, and odor-free substance. Compost can be used for soil conditioning, lawn dressing, and mulching purposes, and as a potting soil component. As a result of using compost as a fertilizer, soil can be improved in the following ways:

(i) Greater resistance to stress conditions such as drought, disease, and toxicity
(ii) Ability to support crops in enhanced uptake of plant nutrients
(iii) Possession of an active nutrient cycling capacity due to the vigorous microbial activity

These advantages ultimately manifest themselves in a reduction of cropping risks, improved yields, and reduced spending on inorganic fertilizers for farmers.

Composting can be undertaken for both wet (green) and dry biodegradable waste types given a sufficient supply. The two waste types may consist of some the common materials listed below:

1. Wet (green) wastes: These waste materials are of high quality since they contain high amounts of nitrogen and moisture. The following are some examples of wet wastes:
   - Food remains and kitchen scraps, including eggshells and stale bread (meat or fat should be avoided)
   - Peelings from fruit and vegetable
   - Freshly cut grass clippings, pruning, tree leaves, weeds, etc.
   - Tea leaves and coffee residues
2. Dry wastes: These wastes have high carbon contents and tend to control the decomposition rate in the composting process. Examples include:
   - Dry grass and tree leaves
   - Tree bark and woodchips
   - Saw dust disposed by timber workshops
   - Straw, maize stalks, etc.
   - Shredded newspaper

The addition of wood ash is encouraged as it acts as a source of some major elements such as potassium calcium, magnesium, etc. Alongside these major biodegradable waste constituents, the addition of the following materials will induce and speed up the process, as they act as catalysts. It is a pre-requisite to have at least one of these materials in small quantities. The catalyst materials include coffee pulp, animal manure (such as that of a chicken, goat, cow, sheep, or rabbit) and also dried blood, bone and fishmeal.

Some materials that are similar to the above but are not suitable for composting include:

1. Charcoal ashes – its high carbon dioxide content interferes with the oxygen supply in the composting system, thus slowing down the process
2. Dog and cat manure – contains harmful pathogens that can disrupt the composting process by affecting the decomposing microorganism population
3. Any organic matter likely to be contaminated with pests or disease
4. Eucalyptus and cassia tree leaves or any biomass suspected to contain substances toxic to microbes
5. Meat and animal fat such as cheese should be avoided as these can attract rodents, raccoons, dogs, cats, flies, and other pests
6. The roots of persistent weeds (such as bindweed, couch grass)
7. Metal, glass, plastic, and nappies

The controlled process of composting is optimized by maintaining environmental conditions in the levels at which microorganisms are thrive. The compost formation rate can be controlled through the composition and through the constituents of the decomposing materials, such as their Carbon/Nitrogen (C/N) ratio, temperature, moisture content, etc.

Microorganisms utilize carbon as an energy source while nitrogen is used for the synthesis of some proteins. Thus, a good C/N ratio is very important for the composting process to be efficient. Compost with an inappropriate C/N ratio can cause adverse effects to both the soil and the plants after its application. The effects of a high C/N ratio can be eliminated by using dehydrated mud, while the effects of a low C/N ratio can be mitigated by adding cellulose.

Moisture content is considered a determining factor in the composting process as microbes need moisture to perform their metabolic functions. Composting material that is too dry may create unfavorable conditions due to a lack of moisture for their activity, while high moisture conditions create anaerobic conditions.

Maintaining optimum temperature is important for the composting process. The temperature should be 50–60°C, with the ideal being 60°C. High temperatures, typically above 75°C, may destroy the necessary microorganisms.

Aeration is also very important as anaerobic conditions cause the death of aerobes, replacing them with anaerobes. The anaerobes can slow down the process while producing odors and highly flammable methane gas. Thus, aeration is undertaken through the frequent turning of the compost piles Viaene *et al.* 2015.

Due to some undesirable conditions, problems may be encountered during the process of composting. One of the main problems is a foul odor, which can result from an excess moisture content in the waste materials, or from compaction. As a solution to this excess moisture, the compost pile can be turned and some dry materials, such as straw, added to it. To avoid compaction, the pile should be frequently turned and decreased its size. Due to an excessive nitrogen content, an odor of ammonia may generate in the compost piles. To avoid this problem, one can add more brown materials, which are rich in carbon.

If the pile is too small and the waste materials have very little moisture, the temperature of the compost pile may drop unintentionally. Poor aeration and cold weather conditions also cause this problem, which can primarily be avoided by turning the pile, increasing the pile size, and insulating the pile with a layer of materials such as straw. As well as this, the temperature of the pile can rise drastically due to a too-large pile size and a higher nitrogen content in the waste materials. This issue can be eliminated by reducing the pile size and turning the piles more frequently; the addition of high carbon (brown) materials can provide better solutions. The presence of meat scraps or fatty food wastes may attract pests such as rats, raccoons, and insects. By removing meat and fatty foods from the pile and covering it with a layer of soil or saw dust, this problem can be avoided. Moreover, one can switch to animal-proof compost bins in order to reduce the risk of attracting pests and animals to the composting area.

A key advantage of the composting process is that its high temperature essentially kills all pathogens and weed seeds that might be found in wastes. Since stabilized compost is no longer subject to sudden chemical changes, it may be safely handled, stored, and used. Mature compost is normally dark brown in color, with an even texture and a pleasant, earthy aroma (Ndegwa and Thompson, 2001).

Various composting techniques are available, each with their own merits and demerits. The major composting methods of composting are as follows:

4.2.3.3.1 Windrow Composting

Windrow composting consists of placing a mixture of raw materials in a long, narrow pile called a windrow. The windrow should be continuously turned and aerated. The turning operation mixes the composting materials and enhances passive aeration. Normally, for dense materials such as manure, the windrows are prepared to a height of 90 cm, where for light and voluminous materials such as leaves, it is constructed to about 360 cm. The width of the piles varies within a range of 300–600 cm. The turned pile is placed in an adjacent space, while a new pile is started in the original area. The process must be conducted continuously, with the piles being frequently fed with composting materials. A free space is required each time the composting pile is turned. Typically, two sites should be prepared: the first for constructing the compost heaps, the second one for turning them. It is important that the land is clear and devoid of any vegetation, with the soil slightly dug in order to loosen it up. If there is any excess water, this will allow the water to drain away.

In the aeration process of windrows, a primarily natural or passive air movement (convection and gaseous diffusion) can be used. The porosity of the prepared windrow influences the rate of air exchange in the waste material. Under the passively aerated windrow method, perforated pipes are embedded through each windrow to supply air for the composting materials. As a result of this mechanism, the need for turning is eliminated. The ends of each pipe are kept open, allowing the air to flow into the pipes and subsequently through the windrows. The hot gases tend to rise up and out of the windrow, dominated by the chimney effect.

4.2.3.3.2 Vermicomposting

The term vermicomposting refers to the application and use of earthworms for the composting of organic residues, in order to derive a humid substance rich with nutrients. Earthworms are capable of consuming all kinds of organic matter to a weight equivalent to their own body weight. The excreta (castings) of those worms is considered rich in nitrate and other available forms of other nutrients such as potassium (K), calcium (Ca), phosphorus (P), and magnesium (Mg). The wastes are eaten, ground, and digested by the earthworms with the help of aerobic and some anaerobic microflora. With the passage of soil through the earth worms' digestive tracts, the growth of bacteria and actinomycetes is induced. Actinomycetes typically thrive in the presence of worms. Therefore, their content in worm casts is more than six times that in the original soil. In this process, biodegradable waste materials are naturally converted into finer, humified, microbially active fecal material (castings). The important plant nutrients are contained in a more soluble form that will be more readily available to plants than to those in the parent compound.

Vermicomposting uses a combination of biological processes, designs, and techniques to systematically and intensively culture large quantities of particular species of earthworms. In the process, these worms speed up the stabilization of organic waste materials.

The produced vermicompost as derived through an accelerated composting process can result in improved chemical, physical, and biological properties and, when applied to the soil, better conditions for plant growth (Jambhhekar, 2002).

4.2.3.3.3 Aerated Static Pile Composting

The aerated static pile method utilizes a piped aeration system or a blower to supply air to the composting materials. The use of the blower allows direct control of the process and the handling of larger piles. The heat and oxygen supply can be easily regulated in this method, which produces high quality compost with improved production efficiency. No turning or agitation of waste materials is required during the composting time period, which typically lasts 3–5 weeks when the adequate conditions are maintained. The pile must be formed properly and facilitate a sufficient air supply that will distribute uniformly through the piles.

Generally, in the aerated static pile technique, the mixture of waste is mounted over a very porous material that acts a base. For this purpose, porous materials such as wood chips and chopped straw are suitable. The porous base material should contain a perforated pipe for aeration. A blower is should be connected to the pipes to facilitate the movement of air, which is either pulled or pushed through the pile.

The initial height of the piles depends on factors such as material porosity, weather conditions, and the reach of the equipment used to build the pile. Generally, the piles are prepared to a height of about 150–245 cm; the extra height may serve an advantage in the extreme cold or during winter time, since it helps to retain heat. The compost pile should be completed by covering it with 15 cm of finished compost or a bulking agent. This finished compost layer is essential to protect the surface of the pile from drying, while also reducing heat loss, avoiding flies, and eliminating potential odor problems within the compost pile due to the filtering of ammonia.

Two common types of aerated static piles are available:

- Individual piles
- Extended piles

Individual piles are long, triangular piles that are about twice as wide as they are tall.. Their width is normally about 300–490 cm, not including the cover. Aeration is undertaken by using pipes, running lengthways, beneath the ridge of the pile. An individual pile may contain either a single large batch of material or a few batches that are generally of a similar age and make-up. For instance, all batches may be composed of three-day-old compost. Individual piles are more practical for when raw materials are available for composting at intervals rather than continuously.

As the pile does not undergo frequent turning, poor air distribution and uneven composting may result. In order to avoid these issues, the raw materials should be carefully selected and mixed well at the initial stage. Maintaining a good structure of the composting pile throughout the entire composting period is important for maintaining appropriate porosity, which will enhance the composting process. In order to maintain the porosity, bulking agents such straw or wood chips can be used in combination with the other waste materials.

The aerated static pile method specifically requires two hollow perforated wooden or plastic rods with diameters of about 2 inches for aerating the pile. These are not generally required for the windrow system (Jambhhekar, 2002).

The advantages of the aerated static pile method are:

- Maintenance of the hygienic standards of the surrounding area, as all the waste materials are kept in one place
- Composting materials are kept away from excess water during rainy periods
- Evenly degraded compost can be obtained due to the regulation of heat loss by insulating the pile

4.2.3.3.4 Rotating Drums

In the rotating drums method, waste materials are mixed, aerated, and moved using a frequently rotating horizontal drum. The drum, which is mounted on large bearings, has a diameter of about 3.35 m and a length of about 36.58 m. It is turned through a bull gear with a residence time of three days. The rotating drum typically supplies capacity for approximately 50 tons of waste materials daily. Due to frequent operation, the composting process starts quickly, providing a swift and enhanced degradation of the raw materials. The highly degradable and oxygen-demanding materials are quickly decomposed using the rotating drum method. A second-stage composting system is generally required for this method, facilitating further decomposition of the raw material. Usually, windrows or aerated static piles can be used for this purpose. In commercial composting systems, it is probable that the composting materials are kept less than one day in the drum, and that the drum is primarily utilized as a mixing device.

During rotation, air is fed from the discharge end of the drum; the material at this end of the drum moves in the opposite direction to the air. Thus, the composting materials that are near to the discharge end are cooled by the fresh airflow. The middle part of the rotating drum receives somewhat warmer air, while the material-loading end receives the warmest air. The composing process is therefore encouraged and enhanced, because the newly loaded and still composting materials, located in the middle part of the drum, receive warm air.

The drum can be fabricated to be either open or partitioned. In the open drum system, the utilized waste materials continuously flow through the drum in the same sequence. The residence time of the waste materials is determined not only by the rotation speed of the drum, but also by the inclination of the axis of rotation. In a partitioned drum, the drum is separated into two or three sections. In each chamber, there is a transfer box fitted with an operable transfer door. This device can be used to manage the ongoing composting process more closely, in comparison to the open drum system. The compartment at the discharge end is frequently emptied by opening its door at the end of each composting routine. The other compartments should also be opened, and the composting materials transferred sequentially. Finally, a new batch of raw material is introduced into the first chamber in order to start the composting process. An important mechanism for inducing the composting process is the retention of a small amount of compost in each chamber to act as an inoculum for the succeeding process. A ridge-like sill, located by each of the transfer doors, facilitates this mechanism by retaining 15%of the previous compost in each chamber for mixing with the new batch. The composted materials are subjected to a screening process directly after exiting the discharging chamber. The larger particles separated by the screening can feed back to the first chamber for further composting (Smith, 1995).

When adapting the rotating drums for a smaller scale, composting concrete mixers, feed mixers, and old cement kilns can be used with some modifications for the composting process. These instruments will also facilitate a rapid start to the composting process through the proper mixing, aeration, and movement of the materials, although they are not as sophisticated as commercial structures.

4.2.3.3.5 In-Vessel Composting

In-vessel composting consists of a group of techniques that confine the composting process to a building, container, or vessel. The composting process is accelerated by forced aeration or the mechanical turning of the composting materials in an in-vessel system. Numerous in-vessel composting methods are available, with each method varying according to vessel combination, aeration devices, and turning mechanisms. When considering these methods, most use some aeration and turning technologies adapted from the windrow and aerated pile methods; they attempt to overcome their deficiencies while also exploiting their attributes.

4.2.3.3.5.1 Bin Composting Bin composting, in which the bin consists of walls and a roof, can be defined as the simplest in-vessel method of composting. The bins are simply prepared using wooden slates and may sometimes be a grain bin or a bulk storage building. These buildings or bins permit a higher stack for composting materials, and allow for the use of floor space more than static piles. Furthermore, the bins can regulate weather issues, eliminate odor problems, and provide better temperature control.

The operations of bin composting is similar to that of the aerated static pile method. Indeed, most of the principles and guidelines applied in the aerated static pile method are suggested in the bin composting method also. Most probably, some means of forced aeration would be incorporated into the floor of the bin. Turning mechanisms are rarely included in the bin composting method, although the addition of an occasional remixing method for the raw materials can enhance and speed up the composting process, allowing better decomposition of the materials (Smith, 1995).

In the method, several bins can be used to transfer the composting materials periodically through a sequence to complete the composting process. Relatively high bins are not recommended since they may result in a greater compaction of raw materials and a higher depth of materials for air to pass through. This would also increase the resistance to airflow, leading to pressure loss. Thus, the operation suggests using a high-pressure blower and raw materials with a strong structure in comparison to that used in aerated static piles.

The advantages of composting in general are:

- Significant reduction of the mass and volume of waste (composting typically results in a 50% reduction in mass and an 80% reduction in volume)
- Stabilization of waste
- Reduction of pathogens present in waste materials
- Improved soil structure through the application of compost as a soil amendment – the soil becomes more "friable", giving it a crumbly texture, and making it beneficial for plant root growth
  - Improved water retention of the soil: in dry climatic conditions this will help to extend plant life, keeping them healthier

- Provides a source of slow-releasing, organic fertilizer for plants
- Economic benefits of compost as an available source of fertilizer
- Population boost of the microorganisms and other beneficial organisms that enhance nutrient uptake and fight plant diseases

REFERENCES

Adams, C.E., Aulenbach, D.B.L., Bollyky, J., Burns, D.E ., Canter, L.W., Crits, G.J., Dahlstrom, D., and Lee, K. *David, H. F. and Liptak, B. G., Wastewater Treatment, Environmental Engineers Handbook*, 2nd ed. CRC Press, 1997, LLC, USA.

Air and waste management association. "Fact sheet: Air pollution control devices for stationary sources," 2007, viewed 27/05/2018, <https://www.awma.org>.

Assamoi, B., and Lawryshyn, Y. "The environmental comparison of landfilling vs. incineration of MSW accounting for waste diversion." *Waste Management* 32(5), 2012: 1019–1030.

Chan, Y.J., Chong, M.F., Law, C.L., and Hassell, D.G. "A review on anaerobic–aerobic treatment of industrial and municipal wastewater." *Chemical Engineering Journal* 155(1), 2009: 1–18.

Davis, M.L., and Cornwell, D.A. *Introduction to Environmental Engineering*, 832–914. McGraw Hill, 4th ed., 2008, United States Environmental Protection Agency, 2000, New York.

De Yuso, A.M., Izquierdo, M.T., Valenciano, R., and Rubio, B. "Toluene and n-hexane adsorption and recovery behavior on activated carbons derived from almond shell wastes." *Fuel Processing Technology* 110, 2013: 1–7.

Environmental Protection Agency, U.S. (EPA), Stationary source control techniques document for fine particulate, EPA Document No. EPA-452/R-97-001. *Office of Air Quality Planning and Standards, Research Triangle Park, NC*, September 30, 1997.

Hegarty, J.M., Rouhbakhsh, S., Warner, J.A., and Warner, J. "A comparison of the effect of conventional and filter vacuum cleaners on airborne house dust mite allergen." *Respiratory Medicine* 89(4), 2000: 279–284.

Henze, M., Harremoes, P., la Cour Jansen, J., and Arvin, E. *Wastewater Treatment: Biological and Chemical Processes*, 1–10. Springer Science and Business Media, 2001, Germany.

Jambhhekar, H. *Vermiculture in India – Online Training Material*, Pune, India, Maharashtra: Agricultural Bioteks, 2002.

Kampa, M., and Castanas, E. "Human health effects of air pollution." *Environmental Pollution* 151(2), 2008: 362–367.

Karpf, R.H. "Basic features of the dry absorption process." *ete.a* 13, 2015: 1–17.

Korres, N.E., Singh, A., Nizami, A.S., and Murphy, J.D. "Is grass biomethane a sustainable transport biofuel?." *Biofuels, Bioproducts, Biorefinery* 4, 2010: 310–325.

Kroll, P.J., and Williamson, P. "Application of dry flue gas scrubbing to hazardous waste incineration." *Journal of the Air Pollution Control Association* 36(11), 2012: 1258–1263.

Lettinga, G., Rebac, S., and Zeeman, G. "Challenge of psychrophilic anaerobic wastewater treatment." *Trends in Biotechnology* 19(9), 2001: 363–370.

Liu,C., Chen, R.H., and Chen, K. "Unsaturated consolidation theory for the prediction of long-term municipal solid waste landfill settlement." *Waste Management and Research: The Journal of the International Solid Wastes and Public Cleansing Association, ISWA* 24(1), 2006: 80–91.

McCray, W.P. "End-of-pipe." In *Science, Zero Waste Institute*, 2011: 148–150.

Metcalf and Eddy. "Wastewater Engineering Treatment and Reuse",In *Wastewater engineering: treatment and reuse.* 4th ed., 2003, McGraw Hill, New York, NY..

Ndegwa, P.M., and Thompson, S.A. "Integrating composting and vermicomposting in the treatment and bioconversion of biosolids." *Biological Research and Technology* 76(2), 2001: 107–112.

Palazzo, L., Woolston, J., and Ristevski, P. *Retrofitting Shaker Baghouses to Cartridge Pulse Jet Technology in the User and Fabric Filtration Equipment – VII.* Proceedings, Toronto, September 12–14, 1994, Air and Waste Management Association, Pittsburgh.

Palm, C.A., Gachengo, C.N., Delve, R.J., Cadisch, G., and Giller, K.E. "Organic inputs for soil fertility management in tropical agroecosystems: Application of an organic resource database." *Agriculture, Ecosystem and Environment* 83(1–2), 2001: 27–42.

Pohland, F., and Kim, J. "In situ anaerobic treatment of landfills for optimum stabilization and biogas production." *Water Science and Technology* 40(8), 1999: 203–210.

Renou, S., Givaudan, J.G., Poulain, S., Dirassouyan, F., and Moulin, P. "Landfill leachate treatment: Review and opportunity." *Journal of Hazardous Materials* 150(3), 2008: 468–493.

Schifftner, K.C., and Hesketh, H.E. *Wet Scrubbers*, 2nd ed. Lancaster: Technomic Publishing, 1996.

Schnelle, Karl B., Jr., and Brown, C.A. *Air Pollution Control Technology Handbook.* Washington, DC: CRC Press, 2002.

Smith, R.C. *Composting Practices*, NDSU Extension Service, North Dakota State University of Agriculture and Applied Science, USDA, 1995.

Sudrajad, A., and Yousof, A.F. "Review of electrostatic precipitator device for reduce of diesel engine particulate matter." *Energy Procedia* 68, 2015: 370–380.

Theodore, L. *Electrostatic Precipitators in Air Pollution Control Equipment Calculations.* John Wiley and Sons, 2008, New Jersy.

Viaene, J., Van Lancker, J. , Vandecasteele, B., Willekens, K., Bijttebier, J., Ruysschaert, G., and Reubens, B. "Opportunities and barriers to on-farm composting and compost application: A case study from northwestern Europe." *Waste Management* 3(1), 2015: 12–25.

Wall, D.K., and Zeiss, C. "Municipal landfill biodegradation and settlement." *Journal of Environmental Engineering* 121(3), 1995: 214–224.

Wang, L.K., Williford, C., and Chen, W. "Fabric filtration (Chapter 2)." In *Air Pollution Control Engineering*, Wang, L.K., Pereira, N.C., Hung, Y.T. (Eds), 59–94. Humana Press, 2004, Totowa, NJ.

5 Cleaner Production

Guttila Yugantha Jayasinghe,
Shehani Sharadha Maheepala, and
Prabuddhi Chathurika Wijekoon

5.1 INTRODUCTION

Sustainable development is an approach that provides longer-term solutions for maintaining commitments and increases the capacity for decision-makers to support their decisions with acceptable information and knowledge (Sirait, 2018). Agenda 21 endorsed cleaner production (CP) as a tool that can contribute to sustainable forms of economic development. The United Nations Environment Program (UNEP) launched the CP technique in 1989 to maintain sustainability in the manufacturing process (Almeida *et al.*, 2010). It acts as a tool for approaching the sustainability vision of an organization.

5.2 HISTORY OF CLEANER PRODUCTION DEVELOPMENT

- 1972 – United Nations Conference on the Human Environment, Stockholm
- 1987 – Brundtland Report (*Our Common Future*) and the concept of "sustainable development"
- 1989 – CP Programme at UNEP
- 1992 – United Nations Conference on Environment and Development and the adoption of Agenda 21
- 1994 – UNIDO–UNEP National Cleaner Production Centers Programme
- 1998 – UNEP's International Declaration on CP

CP facilitates the industrial efficiency, profitability, and competitiveness of organizations while protecting the environment. It is used as a waste preventive approach that is always a better option than waste treatment (Sirait, 2018). CP can be used by any type of organization to make environmental improvements together with significant financial profits. The Pollution Prevention Pays (3P) technique should be adopted by manufacturing companies to minimize their waste and financial losses (Almeida *et al.*, 2010). CP emphasizes pollution prevention as follows:

- Reducing consumption of raw materials
- Reducing energy consumption
- Eliminating or reducing waste generation
- Using optimum techniques in the manufacturing process
- Increasing efficiency and productivity

TABLE 5.1
Different Technical Requirements in CP

Category	Technical Requirements
Products	Life cycle analysis methods Trends in use New products Product lifespan data Product substitution Product applicability
Process	Feedstock substitution Waste minimization assessment procedures Basic unit process data Unit process waste generation assessment methods Materials handling Cleaning, maintenance, and repair
Recycling and reusing	Market availability Infrastructure capabilities New process and product technologies Automated equipment and process Distribution and marketing Management strategies Automation Waste stream segregation On- and off-site reuse opportunities Waste exchange opportunities; closed-loop methods Waste recapture and reuse By-product recovery methods Collection and distribution.

CP is not restricted to industry or to production: it has been successfully applied in the services sector and in municipal programs. Also, it involves a change in attitudes and management practices, the improvement or modification of technologies, and the application of available knowledge (Mahmood, 2015).

CP considers different technical requirements and these are illustrated in Table 5.1.

5.3 ELEMENTS OF CLEANER PRODUCTION

The CP technique consists of four elements:

1. The precautionary approach: polluters must verify that substances or activities do not harm
2. The preventive approach: preventing pollution at the source before it occurs
3. Democratic control: information should be shared with workers, consumers, and communities and the decision-making process should be laded by them

4. Integrated and holistic approach: a life cycle analysis should be conducted to examine all material, energy, and water flows

The application of CP has already enabled the manufacturing sector to reduce pollution risks and provide better management practices associated with wastes (Sobry and Nadzri, 2010). Further, renewed application of the CP approach provides a competitive edge for organizations while providing environmental and economic benefits. CP considers different technical factors such as environmental awareness, consumption demands and limitations, the structure of the economy, processing methods, economic analysis, and process engineering (Mahmood, 2015).

5.4 THE CLEANER PRODUCTION ASSESSMENT APPROACH

Basics steps for cleaner production assessment (CPA) were developed by the United Nations Industrial Development Organization (UNIDO) and are as follows (Sirait, 2018):

1. Planning and organizing CPA
2. Pre-assessment and assessment
3. Evaluation and feasibility assessment of CP opportunities
4. Implementation of identified CP opportunities
5. A plan to continue with CP efforts

5.4.1 Planning and Organization

The first step of CPA consists in the planning and organization of a CP program. Organizations should be aware of the CP process and stakeholders should focus on conducting the CP process in their organization. A small CP team should be formed to conduct the CP program, and this should consist of a cleaner production consultant, different levels of employees, management staff, and company owners. Environmental policies should be reviewed or written. Budgets should be calculated, and total CP program should be designed.

5.4.2 Pre-Assessment and Assessment

All processes should be identified and assessed. Material inputs and outputs, costs, emissions, waste generation, and environmental and health impacts that are associated with the manufacturing process should be reviewed. CP options should be generated based on identified problems. Different tools and techniques can be used to identify the problems and make solutions – specifically, the tools and techniques that are used in the green productivity (GP) approach are also used in the CP process. The assessment stage has two phases: a preliminary assessment and a detailed assessment. The preliminary assessment phase is used to increase understanding about the processes at each site (Berkel and Willems, 1997). The major inputs and outputs are identified and quantified to compare wastes. The detailed assessment phase facilitates the determination of the waste streams of the manufacturing process. Wastes are analyzed by developing a multicriteria waste comparison. CP ideas are generated to directly or indirectly reduce the quantity and toxicity of the analyzed waste streams.

5.4.3 Feasibility Analysis

Each CP option is assessed for its environmental, technological, and economic viability. The final CP options are selected based on this assessment. These selected CP options are then subjected to a feasibility analysis.

5.4.4 Implementation

In this phase, feasible CP options are implemented and monitored. An evaluation program is designed and the organization's culture is changed according to the CP options in order to achieve optimum conditions in the industry. It is a challenge to change the organization's culture, but CP helps to generate waves of innovation that reach beyond the immediate confines of the CP performance objectives.

5.4.5 Continuation

Regular audits are conducted to gather feedback and improve the decision-making process. Progress and gains should be communicated to all stakeholders. Figure 5.1 shows the CP approach.

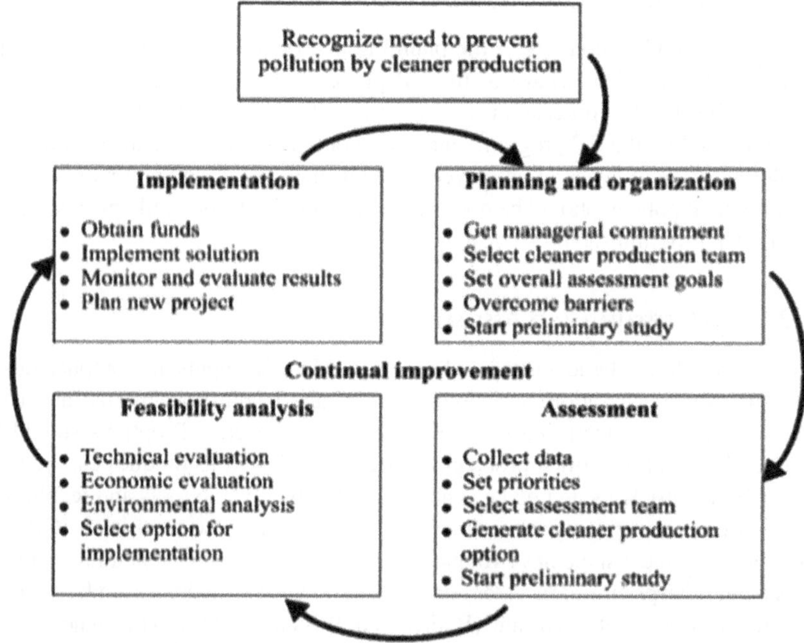

FIGURE 5.1 CP approach.

When organizations use a CP approach it can affect the whole supply chain and encourage others to take action to improve the CP performance in their companies. Leading companies will offer their expertise to their suppliers to support the replication of CP practices (Ashton *et al.*, 2002). However, expanding into the supply chain creates considerable amounts of new tasks within data collection, analysis, normalization, and reporting. Supply chain efforts will take formal shape for many companies that are still in the developmental stages of applying the CP approach (Reddick *et al.*, 2008). Two fundamental principles are applied as follows:

1. The company needs to learn about the internal CP program's implementation process and demonstrate its commitment to it by raising an effective internal operations effort
2. The company should communicate with its suppliers to explain the data requirement, all subsequent commitments, and the benefits of the CP program

5.5 BENEFITS OF A CP PROGRAM

The benefits of implementing a CP program include:

- Improving the energy efficiency of products and services
- Estimating the energy or carbon footprint of products and services
- Motivating companies to develop and market more energy-efficient products
- Gaining higher revenue and profit
- Reducing environmental impacts

CP programs should include all key functions and operating units in an organization. The CP consultant/leader should be involved in the environmental unit, health and safety unit, engineering, sustainability, or operations units of the organization. Effective CP team structures are typically cross-functional and multi-level (Reddick *et al.*, 2008). Most CP investments are small and, when considered individually, can be hard to notice compared to the many larger investment opportunities corporate leaders entertain. CP strategies evaluate and account for the potential for savings opportunities across multiple facilities. The total payback period of CP programs is taken into account and communicated with senior management (Doorasamy, 2015).

Assessment criteria should be developed based on the general CP objectives. The CP approach should be considered at the production level to set targets, and these targets should accomplish the following different outcomes:

1. Minimizing resource consumption
2. Improving the efficiency of resource use
3. Minimizing waste and emissions generated
4. Using the appropriate technologies
5. Minimizing the consumption of hazardous material
6. Implementing environmentally oriented management programs
7. Improving process efficiency
8. Minimizing the generation of hazardous by-products

9. Exploring input substitution and process modification
10. Minimizing the risk of accidents or malfunctions
11. Using energy and energy sources efficiently and consciously
12. Implementing closed-loop systems for materials
13. Integrating renewable sources into the processes
14. Implementing regulatory policy
15. Adopting methodological evaluation tools
16. Adopting environmentally directed, technological solutions through integration with research work

Examples for cleaner production options include:

- Documenting and analyzing consumption based on material and energy flows (Doorasamy, 2015)
- Identifying losses from poor planning and training mistakes using indicators, and subsequently controlling them
- Reusing waste or by-products
- Substituting raw materials and energy (e.g. using renewable materials and energy)
- Improving control and automatization
- Increasing the lifespan of auxiliary materials and process liquids (e.g. avoiding drag-in, drag-out contamination)
- Implementing new, low-waste processes and technologies

Tools used in the CP approach to identify problems are as follows:

- Brainstorming
- Flow charts
- Cause-and-effect diagrams
- Material balances
- Check sheets
- Check lists
- Process flow diagrams
- Control charts
- Spider diagrams

Techniques used in the CP approach include:

- Good housekeeping
- Input substitution
- Better process control
- Equipment modification
- Technology change
- Product modification
- Energy efficiency
- On-site recovery and reuse
- Waste-to-product conversion

5.6 ISSUES WITH THE CP TECHNIQUE

Some issues with the CP approach include:

- Cleaner production is a dynamic concept and difficult to define
- The three aspects of the CP approach (technology, housekeeping, and changing inputs) are problematic to analyze
- Investments in cleaner production must be distinguished from investments in new processes where ecological considerations are not the prime motivation
- Analyses of the investment in clean production only contain the additional investment
- Certain industrial sectors do not lend themselves to the application of clean production
- Analysis of the CP process requires an in-depth knowledge of processes, as cleaner technologies are process-specific
- The timing of investments in CP creates a challenge for the analysis of the size of the market and the employment created (UNIDO, 2010).
- Technological developments will continue to produce CP options

5.7 NATIONAL CLEANER PRODUCTION CENTERS (NCPC)

NCPCs promote CP techniques and investments in CP programs in their countries to maintain sustainability in the manufacturing sector. NCPCs ensure the achievement of organizations' large investments in CP, especially in less developed countries. NCPCs should provide groups of professionals with sound expertise in CP-related matters, and who possess equally robust knowledge of key financial and accounting topics such as capital budgeting, project appraisal, cost accounting, and financing techniques for the implementation of CP programs (Shah, 2008).

NCPCs dedicate more human and financial resources to information distribution, which is essential to create and foster a CP consciousness. Statistics and figures that prove the financial attractiveness of investments in CP, and that guide the choice of alternative technologies available, should be developed; however, databases would be more desirable. NCPCs provide a primary source of information for potential investors in CP technologies and projects (Doorasamy, 2015).

5.8 CP POLICY

CP policies can be developed to maintain the quality and productivity of the CP approach (Queiroz *et al.*, 2015). Also, these policies facilitate the standardization of the CP technique. Basic conditions for CP policy development can be listed as follows:

- Political interest of stakeholders
- Support at the highest level
- Policy continuity and stability

- Environmental awareness and pressure of public opinion
- Broad local ownership of the process
- Integration and mainstreaming of CP into non-environmental policies
- Market conditions that encourage reduction in production costs
- Institutionalization of CP

Different stakeholders who are involved in CP include:

- National governments
- Enterprises and power plants
- Legislative branches
- Environmental service providers and CP consultants
- Sector-specific associations
- The financial sector
- Universities and the education sector
- Non-governmental organizations
- Cleaner Production Centers/CP institutions
- International organizations
- Municipal governments
- Trade workers and laborers

5.9 BARRIERS

There are numerous complex barriers associated with the implementation of CP in an organization (Gavrilescu, 2004; Sirait, 2018), and they can be categorized as follows:

- Lack of resources and capabilities, such as time/staff resources for investigation and implementation, financial resources, required technical skills and knowledge, 'risk-assessments', and investment appraisal competencies
- Lack of awareness and motivation, including an understanding of problems, process controls, product designs and other potential improvements, clear regulatory drivers, clear economic incentives, the true benefits of improvements, competency/skill needs, available tools and techniques, support and its value
- Lack of management culture and structures, such as strategic and holistic thinking, leadership and management commitment, goals and targets, cost-accounting systems, good attitudes to investment and risk, flexible company culture, and openness to change
- Lack of employee involvement
- Inadequate internal and external communication

REFERENCES

Almeida, M.V.B. Biagio, F.G., Donald, H., and Silvia, H.B.. "The roles of cleaner production in the sustainable development of modern societies: An Introduction to this special issue." *Journal of Cleaner Production* 18, 2010: 1–5. doi.10.1016.

Ashton, W., Andres, L., and John, R.E. *Best Practices in Cleaner Production: Promotion and Implementation for Smaller Enterprises*, School of forestry and environmental studies, Yale University, 1–47. Washington, DC, 2002.

Berkel, R.V., and Willems, E. "The relationship between cleaner production and industrial ecology." *Journol of Industrid Ecology* 1(1), 1997: 51–64.

Doorasamy, M. " Identifying environmental and economic benefits of cleaner production in a manufacturing company : A case study of a paper and pulp manufacturing company in KwaZulu-Natal." *Investment Management and Financial Innovations* 12(1), 2015: 235–246.

Gavrilescu, M. "Cleaner production as a tool for sustainable development." *Environmental Engineering and Management* 3(1), 2004: 45–70.

Mahmood, W. "A review on optimistic impact of cleaner production on manufacturingsustainability." *International Journal of Advanced Manufacturing Technology* 18(8), 2015: 1–10.

Queiroz, G., Cobra, R.L.R.B., Guardia, M., Oliveira, J.A., Ometto, A.R. and Esposto, K.F. "The use of lean manufacturing practices in cleaner production: A systematic review." In *5th International Workshop Advances in Cleaner Production*, São Paulo, Brazil, 2015, 1–10.

Reddick, J.F., Blottnitz, H.V., and Kothuis,B. "Cleaner production in the South African Coal mining and processing industry: A Case study investigation." *International Journal of Coal Preparation and Utilization* 28(4), 2008: 224–236. doi.10.1080.

Shah, J. "Cleaner production." *Guidance Notes on Tool for Pollution Management* 2008: 1–10.

Sirait, M. "Cleaner production options for reducing industrial waste: The case of batik industry in Malang, East Java-Indonesia." *IOP Conference Series: Earth and Environmental Science* 106(1), 2018: 1–5. doi.10.1088.

Sobry, C., and Nadzri, W. "Clean Production strategies adoption: A survey on food and beverage manufacturing sector." *Communications of the IBIMA*, 2010, 2010: 1–10.

UNIDO. "Understanding cleaner production." *UNIDO / UNEP Guidance Manual*, 2010: 2–28, Retrieved from http://www.unep.fr/shared/publications/other/WEBx0072xP A/manual_cdrom/G uidance%20Manual/PDF%20versions/Part1.pdf on 21st August 2018.

Index